British Naturalists in Qing China

British Naturalists in Qing China

SCIENCE, EMPIRE, AND
CULTURAL ENCOUNTER

Fa-ti Fan

HARVARD UNIVERSITY PRESS
Cambridge, Massachusetts
London, England

Copyright © 2004 by the President and Fellows of Harvard College
All rights reserved
Printed in the United States of America

Library of Congress Cataloging-in-Publication Data

Fan, Fa-ti
   British naturalists in Qing China : science, empire, and cultural encounter / Fa-ti Fan.
      p.   cm.
   Includes bibliographical references and index.
   ISBN 0-674-01143-0
   1. Natural history—Research—China—History—Qing dynasty, 1644–1912.   2. British—China—History—Qing dynasty, 1644–1912.   3. China—History—Qing dynasty, 1644–1912.   I. Title.

QH51.F36   2003
508'.07'2051—dc21   2003056554

For my parents and
in memory of my grandparents

Contents

	Acknowledgments	*ix*
	Introduction	*1*
I	**THE PORT**	
1	Natural History in a Chinese Entrepôt	*11*
2	Art, Commerce, and Natural History	*40*
II	**THE LAND**	
3	Science and Informal Empire	*61*
4	Sinology and Natural History	*91*
5	Travel and Fieldwork in the Interior	*122*
	Epilogue	*155*
	Appendix: Selected Biographical Notes	*163*
	Abbreviations	*167*
	Notes	*169*
	Index	*231*

Acknowledgments

With a background in science, I came to the United States planning to study cinema but soon discovered the charm of history. Harold Cook, Victor Hilts, and Lynn Nyhart patiently guided me through the intellectual transition in my years at the University of Wisconsin-Madison. Ronald Numbers and Julia Thomas offered timely help at the last stage of the process. They were exemplary teachers, and I am deeply indebted to them for their kindness and good advice. During my years in the lively Midwestern town, I also learned a great deal from cordial conversations with fellow graduate students. I especially thank Tomomi Kinukawa, Yu-fen Ko, Louise Robbins, and Fang-yen Yang, who influenced my research and intellectual outlook in more ways than one.

On moving to Berlin, I was delighted to find myself in a most welcoming and encouraging environment at the Max Planck Institute for the History of Science. It hosted dozens of historians of science, who transformed themselves during weekends into expert backpackers, film buffs, and music lovers. I greatly appreciated the time spent with Matthias Doerries, Karl Hall, Abigail Lustig, Gregg Mitman, Michelle Murphy, Matt Price, Robert Richards, and many others. Friendships made Berlin's dark winter nights as pleasant as its bright summer days. At the institute, I presented draft chapters of the book at the Moral Authority of Nature Workshop, the Materiality Reading Group, the Science and the City Workshop, and a department colloquium. I thank the participants for the many valuable comments I received on such occasions.

Since relocating to the United States, I have been blessed with collegiality and friendships in the History Department and in the Asian and Asian American Studies Program at the State University of New York at Binghamton. My colleagues have been invariably warm, generous, and encouraging. We are a close and energetic community of people who share their interests and lives, far beyond professional concerns. My intellectual and social life in Binghamton has been greatly enriched by the company of Nancy Appelbaum, John Chaffee, Donald Quataert, and many others. Keiko Oda kindly tolerated my retrogressive enthusiasm for Japanese anime and the *Iron Chef* show.

Funding for this project was provided by the Chiang Ching-Kuo Scholarly Exchange Foundation, the National Science Foundation, and the University of Wisconsin-Madison. While I was researching the book, I received considerable assistance from the staffs of the following institutions: the Missouri Botanical Garden, the Botanical Library at Harvard University, the Harvard-Yenching Library, the Arnold Arboretum, the Essex-Peabody Institute, the libraries at the University of Wisconsin-Madison, the Royal Gardens, Kew, the Public Record Office, Kew, the British Library, the Victoria and Albert Museum, the Linnean Society, the Royal Geographic Society, the Royal Horticultural Society, the Royal Pharmaceutical Society, the Natural History Museum of London, the Royal Edinburgh Botanical Garden, the Zoological Society of London, the Entomological Society of London, the Cambridge University Library, the Royal Geological Society, the National Botanic Gardens of Ireland at Glasnevin, and the Public Records Office of Hong Kong. I wish to thank particularly the librarians at Kew Gardens, the late John Thackray at the Natural History Museum of London, and Dr. Shiu Ying Hu at Harvard University for their hospitality and help.

I am grateful to Harold Cook, Lorraine Daston, Victor Hilts, Abigail Lustig, Lynn Nyhart, Robert Richards, and Julia Thomas for reading one or more draft chapters with care and insight.

Thanks are also due to my editor Ann Downer-Hazell for her remarkable patience and editorial skills and to the anonymous reviewers for their detailed suggestions and comments. Distracted by the Sirens of new projects, I left the manuscript in a drawer for longer than I dare to admit. In her quiet and sympathetic way, Ann successfully shamed me into completing the final revisions. Elizabeth Collins expertly copyedited the manuscript.

Earlier versions of Chapters 1, 3, and 4 appeared in *Osiris, British Journal for the History of Science,* and *History of Science.* They have been revised for inclusion in this book.

Although Melissa Zinkin hasn't succeeded in persuading me to read Kant, she has conclusively proved to me that an afternoon walk is a great human invention, especially in her company. She is my most critical reader and gentlest critic. I thank her deeply for her humor, imagination, and loving support.

My interest in natural history originated early in life. I grew up in the countryside, and I still remember vividly that, as a little child, I often followed my grandmother into the hills behind our village to dig bamboo shoots. My grandfather taught me how to catch frogs with a worm dangling from a thin twig—a method that a British naturalist and future Fellow of the Royal Society had found worth reporting to his scientific friends in London. I shall always cherish these memories. Although my dear parents do not really know what I have been working on all these years behind books, their confidence in me has been the most important source of motivation and inspiration. They probably have sacrificed more than they should for their children, who are now scattered in three nations on both sides of the Pacific. Caring, loving, and slightly eccentric, they are wonderful parents. No words can express my deep gratitude to and love for them. This book is dedicated to them.

British Naturalists in Qing China

Introduction

IN SEPTEMBER 1883, an anti-Westerner riot broke out in Canton. The crowd assaulted the foreign settlement, set fire to the houses, and carried away the contents. The residence of the British sinologist Theophilus Sampson was attacked. His herbarium and library burned down. Sampson, ironically, was the headmaster of a Chinese school of foreign languages and had taught many Chinese students. Henry Hance, an authority on Chinese plants and close friend of Sampson, was then Acting-Consul at Canton. Burdened with enormous duties and now extra work, he regretted that he would not be able to pursue botanical study for a long time to come. Hance had developed an easy working relationship with the Chinese Viceroy of the region, who was very fond of flowers. While British and Chinese gunboats patrolled the river, one thousand Chinese troops marched into the district to protect the Westerners. At the time of the riot, Charles Ford, Superintendent of the Hong Kong Botanic Gardens and a friend of both Hance and Sampson, was returning on a Chinese junk from an expedition to the mountains inland. The only foreigner among 150 Chinese passengers, Ford arrived in Canton unmolested. Most of the specimens he had collected would be processed by Chinese hands: except for one recently acquired assistant, the entire staff of his department was Chinese.[1]

As this episode shows, British naturalists in Qing China often found themselves in uncertain, ambivalent, and multifaceted relationships with the Chinese.² They came to a vast empire, a mighty land renowned for its rich flora and fauna since the Jesuit missions of the Counter-Reformation. Numerous unknown treasures, they believed, might be hidden in the interior. Yet their expectations could turn into frustrations at any moment. Human barriers, social as well as official, impeded their scientific inquiries far more seriously than did natural obstacles. It was precisely because of the nature of these difficulties, however, that British investigations of China's natural history had to depend heavily on the Chinese. The naturalists needed connections that transcended national and cultural boundaries. Their enterprise hinged upon Chinese participation.

Although natural history was the most extensive Western scientific research in China, it has been entirely neglected by historians. This book is the first comprehensive study on the subject that is written from the point of view of a modern historian.³ It examines the research of British naturalists in China in relation to the history of natural history, of scientific imperialism, and of Sino-Western relations by tracing the scientific naturalists' activities from a long historical perspective and in a broad cultural context. The book attempts to explain how Western (especially British) naturalists in China and their Chinese "associates" explored, studied, and represented China's natural world in both local and global contexts. At one level, it seeks to reconstruct and explain the purposes, methods, processes, results, and institutional base of the naturalists' research in a China that was beginning to come to grips with Western powers. I examine the boundary drawing and power negotiations among different groups involved in the activities. At another level, the book tries to trace and tease out the strands of knowledge traditions that converged in the scientific representations of the flora, fauna, and geography of China. It exposes the interplay between the discourse of natural history and that of horticulture, of Chinese visual art, of Chinese folk knowledge, and of sinology. My ultimate goal is to explain the formation of scientific practice and knowledge in cultural borderlands during a critical period of Sino-Western relations.

To understand British naturalists in China, it is necessary first to understand the wider culture of natural history in Britain in the eighteenth and nineteenth centuries. During that period, natural history was a science and cultural fashion popular among the public, expressed as a vogue for scientific lectures, botanizing, and collecting insects and fossils, among many other pursuits.[4] The British carried this fascination for natural history with them when they went abroad. Consuls, travelers, missionaries, army officers, and merchants were often also ardent and competent explorers of nature, and many of them kept up an active correspondence with prominent natural historians at home. During this same period, British trade with China grew rapidly, and unrivaled British maritime power made possible transoceanic exploration. British aggression in China increased as the nineteenth century wore on. In spite of their importance, however, neither the popularity of natural history nor the expansion of the British Empire is sufficient to understand the contact between British naturalists and the Chinese that occurred during the research.

In considering the encounters between the naturalists and the Chinese, the concept of "contact zone" or "borderlands" is a valuable heuristic device for defining the space in which the contact took place. Both terms have been used by anthropologists and literary critics to denote the intersecting zone between the temporal and spatial trajectories of peoples of different geographic origins, cultures, and histories. The terms usually do not refer to actual geographic entities, though geography is certainly an important element in them. I have also appropriated for my own use certain concepts in postcolonial studies and cultural studies that emphasize the hybrid and performative aspects of cultural encounter.[5]

All these concepts share at least one perspective. They do not presuppose rigid, inflexible, demarcating cultural boundaries between the parties that came into contact while noting the existence of differences. There were boundaries, of course, but we cannot take them for granted. The task of a historian is to explain why and how the boundaries were formed. Nor do they privilege conventional binary categories such as Chinese/Western culture or civilization in explaining the contacts between the parties. Nor do they, moreover, essentialize power relations. On the contrary, they mark out a space for human actors as agents of historical change. They enable us to see mingling, interaction, accommodation, hybridization, and confluence as well as conflicts across borders of many kinds. Networks of information, for example, often tied Chinese, British, and other Westerners together; and it is neither easy nor helpful to cut them into pieces according to

Chinese/Western oppositions. These concepts allow us to trace the translation, transmission, and generation of knowledge and other cultural productions that did not square with national, cultural, or other conventional categories.

The risk in using these concepts too casually is that they appear to downplay the reality of power differentials. Although the implication is not intrinsic to the concepts, there is still the possibility that they might invoke a rosy picture of free trading of cultural productions. This study increases the emphasis on scientific imperialism in explaining British research after the Opium War (1839–1842); scientific imperialism was correlated with the expansion of British imperialism in China. As an interpretive concept, scientific imperialism stresses the symbiotic, even integral, relationship between scientific and imperialist enterprises. It insists that scientific development and imperial expansion in certain contexts can be best understood as interactive components of a feedback loop.[6] The development of geography, for example, was built partly upon imperial imagination, apparatus, and expansion, which in turn drew upon geographical knowledge for support.

The central focus of current literature on scientific imperialism has been science and imperial domination, so much scholarship tends to dichotomize and essentialize power relations between the West and the Rest. I hope to modify this perspective by exploring how the historical actors negotiated their identities and the boundaries between different cultural traditions. I suggest that unpacking "scientific imperialism"—the conception of which tends to attribute the initiatives and active role to the West, explicitly or implicitly reducing the natives' action to resistance and response—also requires due consideration of the agency of the natives in historical process.[7] Scientific imperialism is a useful concept. It acknowledges the reality of power differentials (in particular contexts) and helps us focus on the purposes, organizations, and ideas of the enterprise of natural history in European expansion. But if we want to know how scientific imperialism unfolded in colonies and other non-Western parts of the world, we cannot ignore the indigenous people, their motivations, and their actions.[8]

In this study, I hope to draw attention to the diversity and range of the Chinese who took part in the naturalists' research. British naturalists had contact with Chinese from all walks of life: officials, merchants, herbalists, compradors, servants, artists, craftsmen of different trades, fishermen, gardeners, Buddhist monks, interpreters, street peddlers, casual laborers, peasants, hunters, and children. Few works have examined the contact between the British (and Westerners in general) and the ordinary Chinese in everyday life. This study, therefore, explores some of the neglected dimensions of Sino-Western relations by using the intercourse between the Chinese and the British naturalists as a point of entry. The historiography in the field of Chinese history seems to have gone through a dialectical process and is beginning to question the doctrine of the China-centered approach, which has dominated the field in the past twenty years or so. A growing interest in maritime China, the Chinese overseas, and Sino-foreign relations does not mean a return to the impact-response or other models practiced by the generation of John K. Fairbank, but rather shows a healthful attempt to understand the history of China in the global context.[9]

Choice of perspective and emphasis is inevitable. This study primarily concerns scientific activities, so it features the British naturalists and their research, leaving room only for the Chinese participating in the activities and for their perceptions and actions in relation to the British. It is true, too, that evidence about the particular Chinese is hard to come by, and conjecture all but unavoidable. Because most of the research activities were private and informal, and because most of the Chinese were socially lower people, one can expect to find only a few traces left by them. These difficulties naturally limit the possibility of telling a truly balanced and multilateral history of cultural encounters. This problem is not unique to the present work, however, as it haunts almost every historian who tries to write about people lost to history.[10] As long as the historian proceeds with earnest caution, the gain will far outweigh the risk.

THE ORGANIZATION of this book reflects the phases of British research into China's natural history and is divided into two parts. Part I covers the time period between the middle of the eighteenth century and the Opium War; Part II traces the subsequent history to the first decade of the twentieth century. The chapters within each part are arranged thematically, though their order is also loosely chronological. They examine the representative and principal activities of British naturalists in China during that time period, and the sequence of the chapters roughly corresponds to the temporal order in which the activities first became prominent.

The British made few noteworthy attempts before the second half of the eighteenth century to investigate the natural history of China. Such exploration grew after the middle of the eighteenth century as the British edged out the other European maritime powers and assumed predominance in the China trade. The increasing British presence in China provided the talent and resources for sustained scientific research. In 1757, the Qing court changed its policy toward Westerners, relegating the trade to one single port, Canton, on the south coast of China. This policy continued for almost a century before it was brought to an abrupt end by the Opium War. Part I discusses British interest and research in natural history in the entrepôt of Canton during this period (commonly referred to by later Western residents in China as "the times of Old Canton"). I argue that the context of the China trade is crucial to our understanding of British research into China's natural history. Not only were most of the people involved in the enterprise directly connected to trade and commerce, but the interests, infrastructure, and practice of the research were integral parts of the social world of the China trade. Chapter 1 describes the naturalists' scientific activities in relation to the commercial and cultural contact in the urban environment of Canton. Having depicted the broad picture, I zoom in on an important episode that exemplifies some of the characteristics of natural historical research in Old Canton. Chapter 2 thus looks at the natural history illustrations collaborated on by the British naturalists and the Chinese artisans for export painting.

Part II examines the period from the opening of the treaty ports in the wake of the Opium War to the collapse of the Qing dynasty in 1911. The political context determined to a significant extent the pattern of British investigations into the flora and fauna of China, because it decided the access the naturalists had to China's natural world. Furthermore, the resources, strategies, and developments of the naturalists' research also depended on, or responded to, the changing Anglo-Chinese relations. I identify three prominent and interrelated dimensions of British inquiries into the natural history of China, namely, the formation of an empire of scientific information, the codevelopment of research in sinology and natural history, and the increasing activity of exploration and fieldwork. The three chapters in Part II deal respectively with these subjects. Drawing upon the notion of informal empire, Chapter 3 maps out the collective profile of the naturalists and the institutions that supported their research. Chapter 4 examines the practice of cross-cultural knowledge translation by connecting the textual tradition in natural history and orientalist scholarship. As the nineteenth century drew to a close, British naturalists gained increasing access to the interior of China. By focusing on scientific fieldwork, Chapter 5 discusses the relationship between natural history and the British imperial culture of hunting, travel, and exploration in China.

Because this book is intended for scholars from different backgrounds, I consistently avoid using scientific names of the plants and animals unless it is absolutely necessary (readers interested in identifying them can refer to the notes). As to the romanization of Chinese terms, I use the *pinyin* system except when the old renditions, like those of certain place-names, are either widely familiar in the West or more convenient for the use of scholars.

I
THE PORT

1

Natural History in a Chinese Entrepôt

NATURAL HISTORY has proved one of the most elusive subjects for historians of science. Even if we limit our attention to the modern period, the enterprise of natural history was still so extensive, so varied in form and practice that it calls for a broad conception of scientific activity, not merely as an intellectual pursuit of an elite circle but as a cultural practice in which many participated. In recent years, scholars have recognized and begun to investigate the historical links between the development of natural history and the expansion of overseas trade and commerce in the early modern world.[1] Organized scientific expeditions were rare until the last decades of the eighteenth century.[2] Before that time, the majority of the exotic specimens of natural history in Europe came either directly or indirectly from trade and commerce. Ships that carried bullion, coffee, sugar, and tobacco across the Atlantic, and those that brought spices, porcelain, tea, and silk around the Cape of Good Hope, frequently had stowed on board exotic animals and new plants purchased from the natives, fossil bones and lizard skins discovered at shops at some foreign port, funny-looking fish caught during the voyage, shells picked up on a beach thousands of miles from home, and countless other curiosities.[3] Some of these items might find their way into the cabinets of a virtuoso, or be mummified in a museum, or be embalmed in the learned volumes of naturalists such as a Buffon or Kircher.[4] With the natural objects came stories, observations, traditions, and pictures about the natural history of other parts of the world, collected here and there, systematically or randomly, which sometimes proved to be as valuable as the specimens themselves.[5]

European traders in non-Western societies did business with the natives—buying and selling goods, establishing commercial networks, and extending social relationships—and they also sometimes collected natural history specimens. More than anyone else, therefore, employees of the European maritime enterprises collected the specimens and observations that fed the scientific communities in Europe and challenged European understanding of the natural world. Hendrik Adriaan van Reede Drakenstein and Georgius Everhardus Rumphius of the Dutch East India Company were exceptional only in the amounts of research they accomplished; their interests in natural history were widely shared by their colleagues.[6]

Trade involved two sides. Like their Western counterparts, natives engaged in overseas trade also acted as cultural agents, collecting information and introducing whatever they thought valuable into their own society.[7] At the same time, they often contributed to Western research into natural history. Studies on natural history and overseas trade that ignore the indigenous people and their knowledge are therefore bound to be incomplete in view.[8] The stream of natural history data mingled with the larger flow of goods and currency, contributing to the cultural exchange and confluence. To some extent, in fact, plants and animals themselves became capital, a means of monetary and social exchange, and acquired multiple kinds of value at once—economic, social, aesthetic, and scientific.

In this chapter, I describe and explain the scientific research of British naturalists in China in the late eighteenth and early nineteenth centuries and its relationship to the China trade. I suggest that the context of the China trade is vital to our understanding of the naturalists' day-to-day scientific activities. Early naturalists in China, most of whom were employees of the English East India Company, depended heavily on their commercial connections with the Chinese for specimens and other scientific data. Even the desire for Chinese plants and animals and the material culture of transporting them cannot be properly understood without the context of the China trade.

The Mysterious Garden

China occupied a special place in the minds of eighteenth-century European natural historians. It was not just a big piece of land inhabited by many short tea drinkers. Jesuit missionaries extolled the cultural achievements and natural riches of the mysterious Middle Kingdom. The tomes of Martini, Kircher, Le Comte, and Du Halde told of its prosperous cities, curious animals, and beautiful plants. Even before 1800, scholars had debated about rhubarb, snakestones, the Chinese musk, and other curiosities allegedly of Chinese origin.[9] Erudite treatises found only limited audiences, however, and most Europeans derived their impressions of China from other sources. One of the most widespread and persistent images of the empire was round-faced people with funny hairdos playing in a garden of beautiful birds and plants. Most Europeans drew this image not from scholarly works, but from paintings on porcelain, furniture, embroidery, and other Chinese goods.[10]

The volume of Chinese goods brought to Europe by the China trade was enormous. Millions of pieces of chinaware—not to mention furniture, tea, silk, paintings, and carvings—had come to Europe via the sea route since the sixteenth century.[11] These material goods and decorative objects had widespread impact on European tastes; *chinoiserie* manifested itself in furniture, wallpaper, and architecture, and blended into the Rococo style.[12] Meanwhile, European industries strove to copy the white, thin, and delicate porcelain.[13]

Westerners' fascination with Chinese aesthetics also influenced the style of European gardens. In late seventeenth-century England, William Temple championed the Chinese idea of irregular landscaping, "sharawadgi," and, in the eighteenth century, efforts were made to translate Jesuit writings on the imperial gardens, Yuan Ming Yuan, as well as other French works on Chinese garden design.[14] The Scottish philosophers Francis Hutcheson and Lord Kames pondered the aesthetics of Chinese gardening.[15] Having actually been to China, the architect William Chambers confidently placed a towering pagoda in Kew Gardens in 1761. Although the English only selectively adopted the features attributed to the Chinese garden and Chambers's reputation actually suffered from his having praised the style of Chinese gardening extravagantly, articles on Chinese gardening remained popular in horticultural magazines well into the mid-nineteenth century.[16] Even more than Chinese gardens themselves, Chinese flowers captured the imagination of the British. Paintings of lavish chrysanthemums, camellias, and tree peonies on scrolls, porcelain, and other trade goods excited curiosity and enthusiasm.[17] New finds brought back by East Indiamen from the famous Canton nurseries, Fa-tee *(Huadi)*, Land of Flowers, added color to English gardens.[18] With their appetites whetted, the British desired even more Chinese flowers and were keen to learn the reputed secrets of Chinese gardening.

Trade goods thus stimulated British interest in China's animals and plants, and British investigations of the natural history of China began with horticulture and garden plants.[19] Until the middle of the nineteenth century, it should be noted, botany and horticulture were not widely divergent enterprises.[20] Founded in 1804, the Horticultural Society boasted of several distinguished botanists among its original members, including Joseph Banks, President of the Royal Society and *ex officio* Director of Kew Gardens, and James Smith, founder of the Linnean Society.[21] Banks, the foremost British botanist of his day, took great interest in pomology and horticulture. His successor at Kew, William Hooker, also managed to dominate both fields throughout his career. Horticulture and related botany occupied a central position in British scientific research in China in the late eighteenth and early nineteenth centuries. This research direction had also been determined by the particular situation in which the British in China found themselves before the Opium War (1839–1842).

Intersecting Worlds, Exotic Goods

Although they entered European trade with China later than the Portuguese and the Dutch, the British had outmatched the other European maritime powers in the China trade by the middle of the eighteenth century. In 1757, the Qing court changed its policy toward Western trade, consigning it to a single port, Canton, at the head of the Pearl River estuary where it meets the South China Sea. By this time, Catholic missionaries had established their networks in some provinces of China, and the Russians had secured a foothold in Beijing.[22] For the maritime powers such as Britain, however, Canton would be their only access to China for the next eighty-five years until the Opium War. During this long period, almost all commercial activities had to go through the Hong merchants (the Chinese licensed to trade with Westerners), and Western traders were confined to a corner of Canton, staying in a row of warehouses, called the Factories, a few hundred yards long.[23] Western residents in China came to refer to this period as the times of Old Canton.

British traders in China consisted of the employees of the English East India Company (which would have the official monopoly of the British-Chinese trade until 1834) and some independent merchants.[24] British residents in Canton outnumbered all the other Westerner residents combined by a wide margin. In 1836, there were about 150 British residents in Canton. The second largest group of Westerners there, the Americans, numbered only about 40.[25] During the trade season, which ran roughly from October to March, Western traders in Canton led a routine life—conducting business, meeting with Chinese merchants, and socializing with visiting traders and captains of East Indiamen. In summer, as required by the Chinese policy, the British merchants withdrew to Macao, where they availed themselves of mansions, yachts, and a leisurely life.[26] In the mid-eighteenth century, tea supplanted silk to become the most important export item.[27] As items for exchange, the East Indiamen brought woolens (which proved to be a failure); European gadgets such as watches; spices, ivory, tortoise shell, and other valuables picked up at Madras, Calcutta, and Malacca; and, most importantly, silver currency. Indian opium and, later, cotton gained importance as the nineteenth century unfolded, though freelance traders, not the Company, pioneered these branches.[28] The Americans became major players in the early opium trade, in part because their specialty, the fur trade, had dwindled.[29]

Canton was a thriving commercial center of three-quarters of a million people, an entrepôt of global importance where goods from different parts of the world were gathered and redistributed.[30] Its maritime trade included the Chinese junk trade, which trafficked in the South China Sea, and the Western trade.[31] The direct overland trade of Canton extended north to the Yangzi provinces, west to Sichuan, and south to Yunnan, covering half of China proper.[32] Curiosities from Europe, dried goods from Southeast Asia, and tropical fruits such as litchi were taken north to be exchanged for silk and other commodities. Merchants from other parts of China also flocked to Canton to trade—and to enjoy the licentious lifestyle for which the city gained notoriety throughout China.

Canton also served as a contact zone for Chinese and Westerners. It is probably true that many of those dwelling within the city walls never saw a "big-nosed barbarian." But much of the population living in the southwestern suburb and elsewhere along the Pearl River directly participated in commercial activities with foreigners. When the East Indiamen arrived at Whampoa, the outer port of Canton, they were immediately surrounded by washerwomen on boats crying out in pidgin for business. Other sampans flocked around. In a good trade season, a mighty fleet of one hundred trade ships could be moored at Whampoa, stretching out for miles. Whampoa's thousands of residents were nearly all engaged in providing different services to the ships (such as labor, repair, and food supplies). Once having landed, foreign traders boarded boats for the ten-mile journey upstream to the city of Canton. They inched along the river, which was always thronged with sampans and houseboats resembling a floating city. In Canton, foreigners could stroll in the neighborhood of the Factories, but they were barred from the walled city, whose gates were always guarded.[33]

Most of the shops on the district's streets specialized in items for foreigners; signboards written in English to attract these visitors could be seen everywhere. During the trade season, the shops swarmed with excited traders and boisterous sailors, from whom, over the years, some of the shopkeepers even received English nicknames. On the eastern side of the district, where the English Factory stood, shops and teahouses on Hog Lane targeted the common seamen, who, after many cups of "samsui," or rice wine, wobbled away tipsily along the street, hands full of presents for sweethearts at home.

By contrast, on the western side of the district, the elegant shops on New China and Old China Streets sought to attract a more refined clientele with an array of fine goods. The numerous carvings, ornaments, and other intricately designed objets d'art proved irresistible to foreign visitors. "I found it necessary to impose some restraint on my inclination to buy," a Yankee confessed.[34] A Briton experienced the same anxiety: "It is almost impossible to see them . . . without feeling tempted to purchase."[35] He also observed that "our workmen have not been able to compete with [the Chinese]" in designing and making these commercial works of art.[36] Visitors also found many painters' studios and apothecaries' stores here.

Both China Streets and Hog Lane ran northward from the river to Thirteen Factory Street, which stretched east to west behind the Factories. On the other side of Thirteen Factory Street lay a maze of narrow, winding streets crowded with sightseers, peddlers, and porters. Each street was devoted to one branch of trade, such as dried foodstuffs (for example, edible bird's nests and sea slugs), cloth and silk, painted glass, and herbal medicines. Carpenter Square, a favorite place of sea captains, was also located in this neighborhood. For generations, mariners engaged the hundreds of Chinese carpenters there to make wooden chests, writing desks, wicker chairs, and other such items for the trade ships.[37]

Every year, foreign seamen and traders visited Canton by the thousands. People from different parts of the world arrived, met, and exchanged goods, information, and other cultural productions; as a result, Canton developed a culture that was at once local and international. Successful Chinese shopkeepers gained reputations among their foreign customers. A local newspaper published by Westerners, for example, described Kheequa, "the celebrated [lacquered-ware] merchant," as "an individual well known of foreigners, and much respected for his integrity, and steady attention to business."[38] Despite the distrustful relationship between Western traders and the local government, throughout the second half of the eighteenth and the early decades of the nineteenth century, the commercial intercourse between Chinese merchants and Western traders continued and expanded with few interruptions.[39] (The brutal version of European imperialism would not manifest itself in China until after the Opium War.) Contact between foreigners and the local Chinese consisted mostly of these commercial relationships. As we shall see, this cultural environment of Old Canton shaped the possibilities and limitations of British natural history in China; scientific research depended on the urban practices in a Chinese entrepôt.

Trader-Naturalists

By the end of the eighteenth century, the Jesuits had been able to gather and transport significant numbers of plants and seeds to France, where they were acclimatized.⁴⁰ By contrast, the British, despite their ascendancy in the China trade, lagged embarrassingly behind France at this time in the investigation of China's natural history. This was due in part to the fact that the British, unlike the Jesuits, had virtually no access to the interior of China; they had seen only a few ports, mainly Amoy and Canton. Until the early years of the nineteenth century, only a few Britons had acquired any acquaintance with the Chinese language or had traveled to the Chinese interior.⁴¹ When, in 1792, Britain finally decided to send its first delegation, the Macartney Embassy, to Beijing to negotiate diplomatic and trading relations, it failed to find among its subjects a single person who knew Chinese. Two Chinese converts at the Catholic school in Naples were hired as interpreters.⁴² This and the other diplomatic mission, the Amherst Embassy (1816–17), would prove to be the sole opportunities for the British to collect natural history specimens in the interior of China without native agency.⁴³ It is not surprising that Joseph Banks seized on the rare opportunities and assigned special gardeners to the delegations for the express purpose of collecting specimens.⁴⁴

Unlike most other places to which European plant collectors went, China possessed a long tradition of sophisticated horticulture that was well regarded, but little understood, in Europe. Accordingly, the instructions Banks wrote for his collectors to China focused on cultivated plants and the importance of acquiring Chinese knowledge of gardening. He ransacked Jesuit writings and compiled lists of important plants and horticultural matters to which the collectors should pay close attention. Among the most important plants were garden flowers (for example, azaleas, the moutan tree-peonies), fruits (for example, litchi, longan), vegetables (if they tasted particularly good), and plants of economic value, such as tea bushes and oak trees. He wanted collectors who came across any plants that were "either useful, curious, or beautiful" to seize the opportunity of procuring them.⁴⁵

In urging collectors to take notes on Chinese gardening, Banks instructed them to gather information about the techniques of dwarfing trees and other modes of manipulating and propagating plants—for, as he explained, "the Chinese are much given to horticulture & are successful Cultivators of [an] abundance of beautiful Flowers."⁴⁶ Of course, the collectors also had to study the far-famed Chinese methods of turning human waste into all-purpose manure, which, once introduced into England, might mend a grievous loss in national productivity. In his brief to Clarke Abel, the naturalist on the Amherst delegation, Banks advised that the gardener accompanying the mission "will never fail of learning something, if he can be brought into contact with his brethren in Pekin."[47] Disappointingly for Banks and his colleagues, the two missions proved to be unsuccessful for both diplomacy and natural history. Still barred from the interior, British natural historians had to make the best of their foothold in Canton.

British sea captains, who often brought plants home with them from Canton, played an instrumental role in Britain's acquisition of Chinese plants. Although a nice new flower could fetch as much as one or two hundred pounds sterling, immediate monetary gain was only a minor consideration. The high casualty rate of the exotic plants in transit rendered their import marginally profitable at best. It was primarily the honor of being the first to introduce a fine flower that captured the captains' excitement and imagination. When a new plant was described in horticultural magazines, the name of the captain who brought it to Britain was always mentioned with appreciation. A new flower also made a great gift. Captain Mayne of the *Atlas*, for example, presented a new chrysanthemum to the Duchess of Dorset for her garden at Knowle.[48] Other plants went to friends, relatives, and superiors or were kept by the captains themselves. Enthusiastic promoters boosted the efforts to introduce ornamental plants from China. Two shareholders of the East India Company, Abraham Hume and Gilbert Slater, mobilized their resources and sponsored the introduction of numerous chrysanthemums and other spectacular garden flowers from China in the late eighteenth century.[49]

Even more than the sea captains, the members of the Canton Factory, owing to the unique position they occupied in Canton, played a significant role in the investigation of China's natural history. Any sea captain or officer could visit nursery gardens in Canton and have plants packed up to send back to Britain. However, even if, against all odds, the plants survived the homeward voyage, this haphazard way of collecting could not satisfy the desires of serious horticulturists or natural historians. Their investigations often required systematic and sustained efforts that only residents in Canton could carry out. The horticultural and botanical establishments in Britain recognized the situation and took steps to build rapport with members of the Factory. Aiding them in this effort was the popularity of natural history and horticulture in Britain, which ensured that many of its subjects abroad possessed at least enough interest and knowledge to be active collectors of natural history data.[50]

The most powerful supporter for the study of China's natural history and a sinophile, Banks wanted to investigate Chinese plants and to collect porcelain for his lady.[51] His duty as custodian of the Royal Botanic Gardens at Kew made the quest for Chinese plants all the more important. To achieve his goals, Banks installed protégés in Canton through his connections with the East India Company and recruited others who were already there. One of these, Alexander Duncan, secured the position of Surgeon to the Factory in 1788 or so through the connection between his brother John Duncan, the incumbent, and Banks and through Banks's influence in the East India Company.[52] John Duncan had been a valuable China correspondent to Banks, and Alexander would continue the work. "I am not . . . insensible, of your exertions, towards the confirmation of my Appointment," Alexander Duncan wrote Banks, whom he promised to supply with plants and information, "as small return for such favor."[53]

In the late eighteenth century, there were approximately twelve regular members of the Canton Factory, a number that would increase to about two dozen in the early nineteenth century. Like the Company's establishments elsewhere, the Canton Factory had fostered a culture of drinking, feasting, and playing rather than that of serious intellectual pursuits. In 1799, George Thomas Staunton, the first British sinologist, came out to join the Factory. Only eighteen when he began his career in Canton, Staunton had already studied basic science and spoke six languages.[54] Staunton soon discovered that his education was better suited for a Georgian savant than a nabob, and he felt miserably misplaced among the resident merrymakers.[55] Both his father, George Leonard Staunton, and his patron, Lord Macartney, were Fellows of the Royal Society and friends of Banks. Well educated and well connected, he became a member of the prestigious Literary Club and a Fellow of the Royal Society at the age of 22.

Like many other educated gentlemen of his time, Staunton was conversant in botany. A friend of his father worried that young Staunton overworked his brains when the man saw him, still a little boy, coming to his garden every day "with a Linnaeus and a Hortus Kewensis in his hand," to study the plants.[56] Although Staunton himself did not systematically investigate China's natural history, his patrician background, intellectual prestige, and interest in botany rendered him a well-qualified patron for Banks's correspondents in Canton.[57] Banks recommended to him a number of promising individuals who either wanted to study Chinese or were interested in natural history; they included two future Fellows of the Royal Society—Robert Morrison, a missionary who would later author the first English-Chinese dictionary, and John Reeves, a Company tea inspector who would distinguish himself as an expert on Chinese plants and animals.[58]

Staunton noticed an improvement in the culture of the Factory about 1810.[59] Several recent arrivals, notably Morrison and Reeves, were intellectually inclined and took an interest in science. Some of them, such as Staunton and the Company surgeons, had already had training in natural history. But even beginners could use the resources and facilities they had as residents in China to collect living plants, specimens, and other scientific data. "Tyro as I am in botany, how can I presume to send any to *you?*" Reeves apologized to Banks.[60] Yet he would go ahead and send innumerable plants to Banks and the Horticultural Society. Similarly, Charles Millet, an official at the Factory, and George Vachell, Chaplain to the Factory, thought themselves to be "mere [collectors] for the benefit of scientific friends."[61] They were, however, major China correspondents of, respectively, William Hooker, Director of Kew Gardens, and John Henslow, Professor of Botany at Cambridge University, in the 1830s.

The naturalists mailed their correspondents and British museums all sorts of curiosities. So little was known about China that the scientific community in Britain welcomed almost anything. Once, for the Museum of the Cambridge Philosophical Society, Vachell packed a single case with, among other items, thirteen packages of dried plants, seven paintings of lizards ("taken from Life" by a Chinese artist), two boxes of insects, ten skins of birds, one bat, fifty geological specimens, two shells, eight litchi fruits, and a small box containing six edible birds' nests and "a skull (of a Chinese Mandarin, beheaded at Canton 1829)."[62] They also received requests for assistance directly from commercial nurseries. The nursery firm Loddiges and Sons asked Morrison to collect for them "fresh ripe seeds, or nuts, of the different kinds of palms [and also] the native or wild trees and shrubs."[63]

The activities of these energetic new arrivals even stimulated some of the senior members who previously had been lethargic in tropical China. After decades of quiet existence in Macao, John Livingstone, a surgeon to the Factory since 1793, became a correspondent of the Horticultural Society and published a burst of noted reports on Chinese gardening in the early 1820s.[64] With this increasing thirst for learning, the library at the Factory quickly expanded.[65] Morrison at one point assembled eight hundred volumes of Chinese medical works, gathered all the drugs found in the apothecaries' shops, and conducted interviews with Chinese doctors with the intention of exploring Chinese medicine and *materia medica*. His friend John Reeves took part in the natural historical aspect of this project. Unfortunately, the enterprise proved to be too ambitious, especially as Morrison was deep in his sinological and missionary work and Reeves knew no Chinese.[66]

By the end of the 1820s, some members of the Factory were taking steps to establish a museum, christened the British Museum in China. George Vachell was named curator, and John Reeves's son, John Russell Reeves, secretary. Morrison and, no doubt, John Reeves were among the promoters of the project. Although the museum would never be completed due to the disbandment of the Factory when the East India Company's trade monopoly ended in 1834, the proposal itself illuminates several points about the naturalists' view of research into China's natural history.

They recognized the necessity of forming a scientific institution to assist their study of natural history. They also understood the "peculiar advantages" they had as residents in China over any of their fellow naturalists in Europe in gathering specimens and information. While they admitted that "China is sealed to us," that obstacle did not seem insurmountable; as resident traders, they could purchase specimens directly from the Chinese themselves. The museum proposal argued that a museum or cabinet of curiosities would give the Chinese a good idea of why Europeans studied nature's objects and what kinds of things they looked for. Such exposure might actually induce the Chinese to take an interest in Western-style natural history. "Considering the taste of the Chinese for live Birds and Fishes, we may hope that the richer Classes may acquire a taste for the same Animals when prepared for the Cabinet." This statement reflects the "civilizing mission" the British saw as underlying such a scientific enterprise. Yet the conception of the museum also rested upon its commercial implications; commerce and science were inseparable in the British scheme. Indeed, the proposal went so far as to claim that "Commerce is the forerunner of modern discovery in Science." The museum, therefore, would contain not only stuffed animals, mounted plants, and cases of minerals, but also examples of Chinese crafts and manufactured goods. The latter would help British traders form "an Opinion as to their power of Competition."[67]

If the function of the planned museum appears to us to have been both commercially aggressive and culturally demonstrative, those involved might have disagreed. The design for the museum was based on the British naturalists' understanding of the progress of civilization, a viewpoint shared by the educated elite in Britain. For them, trade, commerce, useful knowledge (particularly science and the arts), and fair competition were the very cornerstones of a mature civilization.[68] The same concepts had run through much of the political discourse of the British in dealing with the Chinese government during the two embassies, as they would later run through the Opium War.

Gardener-Collectors

The trader-naturalists were residents in China. Many stayed in China for decades, and indeed some became permanent expatriates. With the exception of perhaps two visits home, John Livingstone had been in China for thirty-six years before he died on a trip to Britain in 1829; Thomas Beale, owner of a celebrated menagerie in Macao, spent fifty years there and never returned home. Even Staunton, who hated his life in China, endured a career of eighteen years with only two visits home, one of which was because of his father's death. John Reeves was a member of the Canton Factory for more than twenty years, with only two home leaves. Years or even decades of activity and experience in Canton and Macao enabled these men to establish local networks for collecting specimens and scientific data. They gained the access, logistics, and eventually street-smarts to help them do the research.[69]

By contrast, collectors sent out by individuals and scientific institutes were mostly transient visitors. These Westerners came to China on missions to collect plants of horticultural value. Unlike the trader-naturalists, who pursued natural history and horticulture as a hobby, the collectors were gardeners selected to do the work. Gardening was their trade, and collecting their full-time job in China. China was only one of the many places in the world to which the British sent botanical collectors in the late eighteenth and the early nineteenth centuries. Kew Gardens and the Horticultural Society of London had their plant hunters in Africa, Australia, and the Americas, too.[70]

As visitors, the collectors faced obvious obstacles. Without prior knowledge of China, they could not communicate with the Chinese effectively and could not have the immediate support of their home institutions, so far away.[71] The only way to ensure the efficiency and success of their exertions was to secure the support of local patrons. Except for those in the two diplomatic missions to Beijing, who had the designated naturalists as their direct supervisors, the collectors were all attached to the Canton Factory through the connections between their employers and the East India Company. These arrangements highlighted the somewhat ambiguous status of the gardener-collectors. They were not independent but were instead placed under the protection and direction of their local patrons. As gardeners representing powerful institutions at home, they found themselves in a slippery social position somewhere between gentlemen and servants. Banks warned his collectors against "[taking] upon the Character of Gentlemen."[72] Instead, they should live in the style of servants, in part to save expenses, but also to demonstrate respect and deference to their hosts abroad. On his voyage to China in 1823, John Damper Parks was assigned to the same cabin as the carpenter of the ship. At a time when the whole vessel was made of wood, the carpenter was a highly important figure in the crew of a ship, below only the officers. This arrangement was, therefore, a compliment to the well-connected Horticultural Society, though Parks complained that both the captain and the carpenter treated him coolly.[73]

Even in a foreign land, remote from England, the social hierarchy of English society seldom broke down. During the trade season, the British population in Canton consisted of two major groups. They did not mix, but related to each other according to rigid social rules. Members of the Canton Factory, captains of the East Indiamen, and visiting gentlemen belonged to one category. The large numbers of Jack-tars and laborers, on the other hand, had no place in the Factory. The research activities and achievements of botanical collectors in Canton can be best understood in this social context. As gardeners, collectors were not fully admitted into the club of the gentlemen and, unless they had the help of a local patron, they were left in social limbo, frustrated and isolated. Despite the impressive title of Royal Gardener, William Kerr had "no one . . . to associate with" in the Factory.[74] His local patron, a high-ranking officer at the Factory, had only limited interest in natural history or horticulture, and none of the other Factory members took effective action to remedy the situation.[75] In addition to neglect, according to John Livingstone, who knew Kerr's work well, the small salary Kerr received from Kew Gardens also trapped him in disgrace. While the sum of £100 a year might have been a passable income for a gardener in England, in Western society in China, it was a pittance. Kerr's poverty even cost him his social standing among his Chinese assistants. His personality greatly changed; he took to drink, kept company with "inferior persons," and failed to fulfill his duties as a collector.[76]

Compared with Kerr, the two collectors for the Horticultural Society enjoyed happier situations. Although John Potts and John Parks earned the same meager £100 as Kerr, they had as their patron and host in Canton an ardent naturalist, John Reeves. Reeves closely supervised their work, personally escorted them to the Fa-tee nurseries and the gardens of the Hong merchants, and introduced them to merchant Thomas Beale's Chinese gardener, from whom they gained practical knowledge of Chinese horticulture. Of course, the fact that they stayed in China only for a few months, as opposed to Kerr's eight years, made it easier for their patron to devote consistent attention to their well-being and progress; and due to Reeves's guidance and local connections, their relatively brief time in China was efficiently spent. Parks reported to the Society that "you could not have placed me under a better person than Mr. Reeves, for he has shown me every attention."[77] Besides Reeves, they also received much help from John Livingstone, who had connections with the Horticultural Society, and Beale, who sent his Chinese gardener to places closed to Westerners to collect plants for Potts and Parks.[78]

The encounter between British and Chinese gardeners often began with suspicion and ended, ostensibly, where it began. Professional rivalry characterized their exchange. Whatever they really thought of each other's techniques, however much or little they learned from each other, they did not yield ground easily. The Canton Factory had its own vegetable garden, run by Chinese gardeners, three miles away in an area Westerners could not enter without permission. The plot produced many European vegetables, such as onions. As an expert from Britain, James Main, the collector for Gilbert Slater of the East India Company, was asked by the Factory to teach the Chinese gardeners how to grow good onions. Escorted by Factory members and chaperoned by Chinese officials, Main visited the ground, where the Chinese gardeners, "half a dozen very respectable looking men," waited with all seriousness. Main delivered his instructions. The Chinese gardeners showed disbelief and guffawed. They doubted that the European method would work in China.[79]

John Parks had the opportunity to observe closely the practice of Thomas Beale's gardener, who was, according to his employer, one of the best Chinese gardeners in Macao. He always provided a reason or a good explanation for whatever he did. Beale also told Parks that "the only way to please a Chinese gardener was to let him do as he pleased," particularly as regards the methods "they took much pride in." They "thought there was none in the world to equal themselves in the management of Chrysanthemums." Parks could hardly swallow this. He opined that the Chinese were "sadly behind the Europeans" in their methods of gardening. "[Chinese gardeners] were the most slovenly fellows," he sneered. "In England, they would be considered no Gardeners at all." All the same, he noted that the method Beale's gardener adopted to propagate plants from cuttings had had "a wonderfully good effect." Beale's gardener, for his part, resisted the advice of Parks and Beale on changing the compost mixture for camellias. "Only with great difficulty" did the combined British forces prevail.[80]

In Canton, gardener-collectors and trader-naturalists collected new plants, curiosities, and other scientific data and transported them back to Britain. Due to their different backgrounds and roles in the enterprise, they filled different capacities and pursued the activity accordingly. Their intentions also differed; one's hobby was the other's occupation. Yet they shared by and large the same sources of specimens and scientific data: their fieldwork sites were the same gardens and marketplaces.

At the Markets

With only minimal freedom of movement, the naturalists rarely went farther on their "expeditions" than the gardens, nurseries, fish markets, drugstores, and curio-shops in the district. The collectors for the Horticultural Society, for example, made most of their collections from local gardens and nurseries, supplemented by the fruits of a few excursions to the hills in Macao. Everything else came from the Chinese collectors.[81] Although collecting specimens at the markets might appear idiosyncratic to a modern reader, it was actually neither new nor unusual. In Europe, natural historians of the Renaissance and early modern period routinely collected in these locations.[82] Until the nineteenth century, exotic animals brought to Europe had mostly been purchased from the natives, not procured through any arduous, heroic actions of European naturalists. For British naturalists in Old Canton, their fieldwork sites were the markets. The vendors in the few streets in the neighborhood of the Factories and the Fa-tee nurseries not far upstream supplied the bulk of the specimens the naturalists sent home.[83]

At the markets, the Chinese sold animals and plants to other Chinese as well as to Westerners.[84] Like Europeans, the Chinese kept domestic animals—livestock and pets. Birds were their favorite pets; the Chinese reared songbirds in cages, training their voices, taking them for airings in the morning, and feeding them with choice food.[85] In addition to birds, the vendors sold many other creatures in the markets. Chickens, ducks, and geese abounded.[86] "Pails and buckets containing live fish may often be seen, ranged by the side of the baskets of vegetables and living animals."[87] The Cantonese kept many varieties of dogs, but, according to a British witness in the 1830s, those found in the streets seemed to be of the same kind. They were used as guard dogs, not lap pets. In the markets were little puppies in cages and baskets for sale. Vendors also sold rats, usually skinned, as a source of meat for the poor. Wildcats were carried from the countryside to the markets to be sold as game.

Western sailors often picked up curious animals at ports around the world for company on long boring voyages or to sell later for profit.[88] When Western seamen arrived in Whampoa and Canton, they were greeted by animals on display for sale. A French visitor in the late eighteenth century came upon an "ourang outang," whose human-like face and sad expression struck him; and but for the foreseeable difficulty of taking care of it on the long voyage home, he would have bought it. This Frenchman did buy a wildcat and took it aboard ship; the cat soon escaped, helped itself to some chickens, and then leaped overboard and swam away down the Pearl River.[89]

The Cantonese even sold insects, living and dead. According to the local histories, large spectacular butterflies were brought from the Lofu Mountain, sixty miles away, to be sold in the streets of Canton.[90] They were tethered by threads to thin bamboo sticks, so that they could flutter and display their beautiful wings. We do not know if the custom persisted in the late eighteenth and early nineteenth centuries; no Western account records it. However, even Westerners with lazy legs had no difficulty obtaining specimens of Chinese insects, because boxes of them were sold in the shops about the Factories. Carefully preserved butterflies, beetles, dragonflies, and other insects were placed in camphor-wood boxes; the bodies of the bigger insects had been "hollowed and stuffed with an aromatic powder."[91] An entomologist in Britain stated that boxed Chinese insects were sent in great numbers to Britain.[92] To be sure, the Chinese had no idea of Western entomology; they collected and arranged the insects according to their brilliant coloration. Despite the erratic juxtaposition of kinds, these specimens gave Western naturalists some basic ideas concerning China's insects.

Ever sensitive to the market demand, the Cantonese tradesmen also made available to the Westerners "live birds and [sea] shells."[93] Most of the hundreds of birds kept in Thomas Beale's aviary in Macao had been purchased from the Chinese.[94] If the customer wanted special birds from other parts of China, the Chinese tradesmen could make arrangements to procure them through their networks of inland trade. Beale succeeded in acquiring birds from the Yangzi valley, Yunnan (a province in southwestern China, near Tibet), and even the region of Tartary, thousands of miles north, though, in some cases, it took years of patient waiting.[95]

British visitors to Canton took advantage of this ready supply of domestic animals and zoological curios. The dish-faced Chinese pig arrived in England during this period, having been "brought hither as sea-stock, or otherwise."[96] It added new blood to the British livestock and thus contributed to agricultural improvement. Many of the dozens of living Chinese animals donated to the Zoological Society of London in the 1830s must also have been picked up in the markets of Canton. They included dogs, geese, mandarin ducks, pheasants, tortoises, and other animals common in the Cantonese markets.[97] Specimens of small mammals, fish, and birds that John Reeves, his son, and George Vachell sent back to British scientific institutions were obtained by the same means.[98]

However, the most important fieldwork site for the naturalists was the Fa-tee nurseries, three miles upstream from the foreign Factories. Fa-tee was one of the several flower markets of Canton; the other ones, however, were out of Westerners' reach. During the Ming dynasty (1368–1644), the Chinese developed a widespread culture of flowers and gardening. Flower fanciers busily bred new hybrids, and nurseries flourished in the outskirts of big cities. The essayist Zhang Dai (1597–1689) tells us that a variety of the tree peony was cultivated by the acres like a crop in Yanzhou, northern China.[99] Huatian ("Flower Fields"), a village a few miles west of Canton, was entirely devoted to planting jasmine. A portion of the production was probably used in scenting tea; but there were places in Canton, including Huashi ("Flower Market"), where jasmine and other flowers were sold in large quantities.[100] Fa-tee was one of several similar locations in the Canton area, though it distinguished itself from the others by its many nurseries and wide variety of flowers cultivated.

Fa-tee excited the admiration and amazement of European visitors in the eighteenth century, and it became "customary" for traders and the officers of the East Indiamen to visit famed gardens while they were in Canton.[101] The nurseries delivered cut flowers, rented out potted flowers for celebrations, and sold seeds and plants of numerous kinds. Chrysanthemums, orchids, tree peonies, dwarf trees, citrus and other fruit trees, camellias, azaleas, and numerous other ornamental plants in pots were displayed for sale. Many of them were tropical plants or southern varieties of flowers common in China proper. When Shen Fu, a connoisseur of flowers from Suzhou, in the lower Yangzi region, visited Fa-tee in the late eighteenth century, he was surprised to find that three or four out of every ten flowers were new to him.[102] During the Chinese New Year festivities, the flower markets boomed. "The *florimania* is even more prevalent in China than in Europe," a gardener-collector observed. "One hundred dollars is freely given for fine specimens of favorite plants," such as the ink orchid, "which is not at all an uncommon plant!"[103] The nurserymen even took the trouble to ship down budding moutan peonies and some other flowers from the Yangzi region, several hundreds of miles north, every year, as these plants did not flower in tropical Canton.[104]

Western visitors and residents in Canton liked to visit Fa-tee for fresh scenes and new plants. To show its leniency toward foreigners, the local Chinese government gave special permission to accommodate this need.[105] This gesture extended to allowing Westerners in Canton to celebrate Chinese New Year at Fa-tee. Their merry picnics, lubricated with wine and songs, amused many Chinese onlookers.[106] The nurserymen at Fa-tee were used to having foreign customers. Some of them, such as Old Samay in the late eighteenth century and Aching in the 1820s and 1830s, regularly supplied flowers to the Factories.[107] Originally a flower market for the Chinese gentry, Fa-tee made adjustments for its Western customers. Aching, for instance, put out a shingle painted in English: "Aching has for sale, fruit trees, flowering plants, and seeds of all kinds."[108] At Fa-tee, large quantities of seeds, "neatly folded in showy yellow paper," were sold to Westerners.[109] Upon examining them, James Main, a trained gardener, believed that they were terribly overpriced. He was probably right, but that only proved how badly Western customers wanted Chinese plants. It was a seller's market. British visitors in Canton also bought large numbers of living plants and tried to transport them back to Britain. Due to technical difficulties, most plants perished en route.

Yet the accumulation of the surviving plants changed the relative positions of British customers and Fa-tee nurserymen. By the 1830s, British visitors to Fa-tee more often than not expressed disappointment. They claimed that they saw few new things. Moreover, the rapid development of British horticulture and the expansion of the British Empire during this period enabled them to possess ornamental flowers from all parts of the world. British flower lovers were now harder to please. John Reeves, who knew Fa-tee as well as any Westerner did, lamented the decline of Fa-tee that he witnessed in the late 1820s. He criticized the gardeners for not trying to collect and cultivate wild plants, "of which they have so many beautiful ones." "The gardens at [Fa-tee] are falling off fast."

Perhaps Reeves was being unjust. He wished that the Chinese nurserymen had done more for their British customers, cultivating new ornamental plants that had not yet been "brought to England."[110] Yet the nurseries probably would not have survived on Western demand alone, and what the British thought beautiful might not have fit the taste of the Chinese gentry, whose partiality for the plum tree, the chrysanthemum, and other plants full of cultural symbolism was hard for Westerners to comprehend.[111] The Fa-tee gardeners might have reasoned that a drastic reform to meet the demands of Western customers would not pay off in the end, and so they might have chosen not to pander to that sector of the market. Whatever caused the decline of Fa-tee, the changing British perceptions of the place signaled their now uncompromising confidence in their own (horti)culture and their increasing knowledge of Chinese plants.

Westerners frequented the marketplaces in Canton to hunt for new animals and plants. The naturalists, the regular traders and sea captains, and the Chinese traded in animals and plants for very different reasons, but the mechanism of the China trade provided the common ground. It was part of the international commerce that also circulated tea, silk, porcelain, and export art. In addition to visiting the local markets, the naturalists also secured a supply of specimens through broadening the already-established friendships, commercial relationships, and similar modes of exchange with the Chinese. Horticulture and natural history actually formed part of the gift relations in the commercial world in Old Canton. The artisans, shopkeepers, and gardeners in Canton were not the only Chinese from whose knowledge and resources the British naturalists benefited. The Hong merchants, higher up on the social scale and the approximate counterparts of Western traders in China, also proved helpful allies in the naturalists' quest.

In the Gardens of the Hong Merchants

The Chinese government licensed about a dozen Hongs, or trading houses, to manage the trade with Westerners. The Hongs were family-owned and usually passed down from father to son. Although the Hong merchants were constantly harassed by the government officials and many of them were insolvent, a few did extremely well. The Hong merchants had regular contact with Westerners, conversed with them in pidgin, and were more or less familiar with Western customs. The more prominent ones were highly respected by their Western trading partners for their credibility, wealth, generosity, and business acumen. Indeed, Max Weber was so puzzled by the discrepancies between his theory of capitalism and the praise offered to the Hong merchants by Western traders that he could do no better than to assert that the Hong merchants had learned conduct from Westerners in Canton.[112] Modern scholars, on the other hand, tend to place these merchants within the context of the commercial culture of Ming and Qing China. Like others in maritime China, the Hong merchants were both flexible and traditional. While they longed for the honors of mandarins, purchasing official titles and emulating their sophisticated tastes, they also possessed the worldliness, adaptability, knowledge, and shrewdness of good players in the arena of international trade.[113]

For example, Puankequa, the premier Hong merchant in the late eighteenth century, had traded in the Philippines three times before he joined the Hong business. He spoke foreign languages and was highly attuned to the social civilities of international trade.[114] His famous banquets for Western merchants and sea captains came in a sequence of two. The first meal was served in the Western style, knives and forks for everyone, even the Chinese, who wielded them awkwardly. The banquet on the next day was in the Chinese style, including chopsticks. Wines, Western and Chinese, flowed freely. Entertainment followed the banquets. On the first evening, a Chinese opera was performed, full of fighting, dancing, and loud music, and spiced up with characters in Western dress. On the second evening, the guests enjoyed a series of shows of fireworks, acrobatics, and juggling in their host's elegant garden.[115]

Most Hong merchant families in Canton shared Puankequa's sense of hospitality and could accommodate Western manners. The social intercourse between Hong merchants and Westerners hinged on, but was not limited to, commercial relationships. The amiable Consequa, a Hong merchant, was much liked by Westerners in Canton; his final bankruptcy indeed may be attributed in part to a too easygoing style of dealing with Western merchants.[116] One of Puankequa II's brothers, a retired scholar-official nicknamed "the Squire," was "very partial to the society of the English."[117] A young American lady met old Mowqua II in Macao and found him "a great character." "He was very gallant, I assure you," she told her sister in Salem, Massachusetts.[118] A major British trader admitted that Puankequa II was "a clever able man," and he confessed that "I like better to dine than to do business with him."[119] The serious Howqua II annoyed some Western traders because of his tendency to do business only with particular firms, but his intelligence and soberness commanded respect.[120] Young Minqua, who was "remarkable for his polished manners," learned to play Western card games and "became quite a celebrity in the foreign community."[121]

Some Hong merchants, notably Punkequa II and Howqua II, were immensely rich, and even Western traders in Canton, who handled shiploads of expensive goods day in and day out, were awed by their wealth.[122] Their villas and gardens on the island of Honam (Henan), across a river from the Factories, charmed Western visitors. The gardens of Chowqua and Consequa, also on Honam, were no less elegant. These and other Hong merchants' gardens were in the Chinese style, which consisted of many smaller grounds, divided by dwellings, gates, or low fences. Narrow paved paths connected the houses. Thousands of potted plants, containing such Chinese favorites as chrysanthemums, camellias, and dwarf trees, were arranged on the terraces, along the walks, and around the pavilions. Miniature mountains or craggy rocks rose out of the lily ponds, in which thrived fish, water lilies, and tortoises. Small bridges arched over the water. Deer, cranes, peacocks, and mandarin ducks added beauty and animation to the picturesque scene.[123]

Like their European counterparts, the Chinese gentry had also developed and shared a widespread culture of flowers and gardens.[124] During the Qing dynasty, some gardens were famous for the refined taste reflected in their designs. The poet Yuan Mei's garden, for example, was renowned for its testimony to the owner's unmatched artistic accomplishments. The elaborate gardens of the salt merchants in Yangzhou, a Lower Yangzi city, occupied vast grounds and extended for miles on both banks of a river, as vividly described in Li Dou's *Yangzhou huafang lu* (c. 1795); as in Europe, such displays were as much an exhibition of wealth as a pretension to cultural status.[125] In Canton, the Hong merchants' gardens, impressive though they were, were still overshadowed by the best ones at Lizhiwan, in the western suburb of Canton.[126] Fine gardens were only part of an opulent merchant's aspirations for cultural elevation. Among the Hong merchants, the Puankequa and Howqua families were major art and book collectors in Lingnan or the province of Guangdong. Puankequa II himself dabbled in poetry, and there were several more literary aspirants among his family.[127]

Westerners in Canton frequented the Hong merchants' gardens in Honam both on formal occasions and during daily walks; these visitors "were at any time civilly admitted by the servants in charge."[128] The gardens contained rare flowers not found in commercial nurseries, such as fine varieties of moutan peonies. The moutan was one of the plants the British most desired, in part because it was a native of temperate climates and would presumably do well in Britain. But precisely because of this nature, the plant was rare in Canton.[129]

Urged by Joseph Banks, the Duncan brothers eagerly looked for the moutan during the closing decades of the eighteenth century. They obtained the plants from the Hong merchants and even from the Hoppo, the chief customs officer, who was generally disliked by the traders.[130] Other naturalists, too, sought plants in the Chinese merchants' gardens. Through the introduction of the East India Company's members, James Main gained access to the gardens of Munqua and of Shykinqua II.[131] William Kerr benefited mostly from Puankequa II among the Hong merchants.[132] Puankequa II actually exchanged letters and gifts with Joseph Banks through the medium of the Canton Factory. He also sent Banks rare plants, including a very old dwarf tree and many pots of the finest moutans.[133] Within a few months of his arrival in China in 1812, Reeves had already dined two or three times at the Squire's and found a treasure in the two to three thousand pots of fine chrysanthemums in the latter's garden.[134] The second day of his sojourn in Canton in 1821, John Potts was taken by Reeves to the Squire's garden, and in the next few days, they visited several more Chinese gardens.[135]

All these activities constituted part of the social world of the traders in Canton. The courtesy of gift exchange continued from the regular commercial favors, and animals and plants were among the welcomed gift items.[136] When Puankequa II asked an American friend for emergency medical help, he sent with his invitation a precious pheasant.[137] Similarly, a shared interest in horticulture likely strengthened the friendships and commercial connections. Innumerable new plants were introduced into Britain through the trader-naturalists and the Hong merchants. Indeed British naturalists at home were so impressed with Consequa's generous offers that they named a plant after him.[138]

The traffic of plants actually went both ways. Thomas Beale distributed several kinds of magnolias among Chinese merchants in Canton.[139] Those who intended to collect plants in China, including sea captains and botanical collectors, commonly brought plants out with them to exchange for Chinese specimens. Influenced by the Enlightenment idea of the scientific commonweal, Banks suggested that Clarke Abel take with him some lemon trees because Banks had heard that "the Lemon is not known in the Chinese Gardens."[140] Most of the plants brought from Europe were probably first destined for the gardens of Westerners in China; but if they were appreciated by the Chinese for their beauty, novelty, or utility, they would surely find their way into Chinese gardens.[141]

The Hong merchants' knowledge about Chinese animals and plants in general and horticulture in particular was probably limited. According to Reeves, who stayed in China in the 1810s and 1820s, the Squire was the only one in the Chinese circle who paid "decided attention to flowers."[142] Still, as members of the Chinese gentry, the Hong merchants were by no means indifferent to horticultural matters. The Duncan brothers constantly received assistance from the Hong merchants and the local gardeners in answering Joseph Bank's queries. Banks sent Alexander Duncan a book of Chinese plant drawings, and Duncan asked "all the literati of Canton," including the Hong merchants, for the Chinese names of the plants.[143] As regards the moutan, he had "a long conversation" with Puankequa II, who explained the nature and habits of that flower.[144] Some of the investigations concerned economic botany. John Duncan inquired about the method of preparing hemp.[145] Alexander engaged a Chinese to collect samples of the tree whose sap was used to make lacquer.[146]

It is true that a zealous naturalist like John Reeves might well have exhausted the botanical lore of his Chinese friends. It is probably true, too, that his mania for animals and plants might have left them baffled. Even some of Reeves's British friends in Canton seem to have had reservations about this natural history business, if we can judge from their decision not to subscribe to the British Museum in China.[147] What is certain is that much horticultural knowledge and many specimens were exchanged in the delightful gardens of the Hong merchants. When Western traders interested in natural history and horticulture visited the Chinese merchants' estates, strolled in their gardens, and examined their flowers, the conversation between the foreign visitors and their Chinese hosts, in the hybrid tone of pidgin, ranged from the price of tea to the modes of cultivating tree peonies.

Transporting Plants

As in collecting, transporting specimens back to Britain depended on the existing infrastructure of the China trade. Dried seeds, bulbs, living plants, pets, and livestock were all taken aboard along with tea, silk, porcelain, rhubarb, cassia bark, and other Chinese goods. The northeastern monsoon began in November, and some of the East Indiamen left by the year's end, loaded with the prized first spring tea for the eager British market. On the homeward voyage, most ships touched only at the Cape or, later, St. Helena, before nearing British waters, a voyage lasting four months or more.[148] Keeping exotic plants alive on such a long voyage was not a trivial matter left to seasick gardeners; it was a serious problem that also challenged many a learned savant.

Live plants as cargo were at the mercy of the ship's captain and the crew. Some captains were themselves plant lovers and routinely carried their own horticultural souvenirs; they tended to take better care of others' plants as well. Understandably, the naturalists preferred to entrust their plants to these captains whenever possible. Otherwise, the plants that they had taken so much trouble to procure might be neglected once the ship put to sea. Alexander Duncan was "much pleased" to have obtained Captain Wilson's promise to transport plants for him, as he knew that "that Gentleman has a botanical turn."[149] Captain Biden's offer to do likewise for John Reeves and John Potts, on the other hand, caused them some embarrassment. The well-meaning captain knew little about plants, so after careful consideration, Reeves and Potts decided "not to send [their] best plants by his ship." Instead, they would let him take only "a case of wild plants of the hardiest kind."[150]

Caring for living plants during a long voyage was a demanding job. The plants were usually placed in wooden boxes or various kinds of plant cabinets. Throughout the journey, they endured harsh conditions—salt spray, lack of water, scorching sun or insufficient light, and fluctuating temperatures. The casualty rate was extremely high. The best spot for plants was on the poop deck, where they enjoyed plenty of light but avoided the worst splashes of seawater. Often this prized spot was taken by a grandee's plants, and the less powerful had to pile theirs in corners of the ship. Even if one's plants were granted a good spot, their fate was by no means secure. Large numbers of plants put up high could tip the balance of a ship in a rough sea or cause other accidents, so they were usually the first cargo jettisoned in foul weather. Plants were often lost in accidents, one way or another. Clarke Abel, William Kerr, and John Potts all had such devastating experiences.[151] In seconds, the fruits of many months of labor could be crushed or swept into the sea.

Living plants required a lot of fresh water, a precious item for the crew, next only to liquor and beer. Because the quota was barely enough for the crew's needs, the person in charge of the plants had to be frugal with every drop of the water set aside for them. When it rained, he would expose the plants for a soaking and collect rainwater for future use. Salt spray was a fearsome killer of the plants; salt quickly crystallized on the leaves and had to be wiped off with a sponge, sometimes several times a day. Animals on board posed another problem for the protection of the plants. Dogs, cats, pigs, goats, and monkeys—all of which were common on every ship—could cause havoc. Moreover, because the ship had to travel across different climatic zones, only hardy plants were likely to survive. In unstable weather, the plant cabins or boxes had to be opened and closed many times a day. Even packed seeds and bulbs were not safe. Slight oversight invited rats to munch on them; worms ate them. Oily seeds spoiled easily in hot weather. If living plants survived all the trials in the first stage of the voyage and arrived at St. Helena safe, they were planted in the gardens on the island for two months or so to recuperate. Other ships bound for Britain then loaded the plants to complete the last leg of the long, arduous journey.[152]

This hazard-filled method of transporting living plants from China continued until the Wardian case, which was basically a sealed miniature greenhouse, was widely adopted after the early 1840s.[153] Grumbling sailors could hardly be made to take good care of the plants, so rewards were offered. Kew Gardens would pay five guineas to the seaman selected to oversee Kerr's plants if he did a good job.[154] Captain Wilson, who carried out "a large collection [of European plants] to exchange for Chinese Plants," engaged an English gardener to look over them.[155] Chinese gardeners occasionally went abroad with Chinese plants. Following the instruction of Joseph Banks, William Kerr sent a Chinese boy brought up as a gardener, Au Hey, on board to take care of the plants he shipped to Kew Gardens. While at Kew, the young gardener greatly amused Banks and others.[156]

Almost all plants from China were shipped to Britain by the route described here. To a great extent, the material culture and infrastructure of transportation determined the kinds of Chinese plants that naturalists in Europe received. Reeves sent certain kinds of azaleas by the hundreds to his correspondents in England; not a single one survived the journey.[157] If the plants and seeds successfully made their way to Britain, they became valued properties of the Horticultural Society, Kew Gardens, commercial nurseries, or whatever institution had financed the project. As with Chinese export art, plants were exotic goods. In 1833 Captain M'Gilligan's double azalea, the first of its kind ever brought to England, sold for £100 to Thomas Knight's Exotic Nursery, and people thought that he was "much abused for letting [Knight] have them so cheap."[158] The Horticultural Society had offered an award of £250 for the first living specimen, though the deal finally fell through.[159]

A price of a few hundred pounds seemed reasonable, given how much labor was necessary to obtain these plants and how desirable Chinese roses, camellias, asters, azaleas, lilies, chrysanthemums, and tree peonies were on the English horticultural market. Based on his twenty-five years' experience, John Livingstone estimated in 1819 that only one out of every thousand plants shipped from China made its way to Britain, which meant each plant successfully transported cost the enormous sum of £300. Plants themselves were cheap in Canton, six or seven shillings each on average, wooden chests included. It was the transoceanic shipping and the high death toll that made acquiring them so expensive.[160] If the plants arrived in good condition, the introduced flowers would be acclimatized, named, and placed on the market. The choicest varieties would first be presented to patrons and dignitaries and be named after them, such as the Lady Banks rose and the Lady Hume's Blush.[161] Other, less fancy ones would also receive English names. In this way, exotic flowers from China became regular inhabitants of English gardens, rubbing shoulders with fellow emigrants from Australia, Africa, India, and the Americas—a testimony to British maritime power.

HORTICULTURE and natural history formed part of the circulation of aesthetics, information, wealth, goods, and other material and cultural productions in global maritime trade. The urban environment of Canton turned out to be ideal for this area of research. Behind the hustle and bustle of a trading city lay a fertile field of scientific inquiries. The shops, the gardens, and the markets; the artisans, the nurserymen, the barefoot street vendors, and the silk-robed merchants; the well-oiled wheel of social intercourse and commercial transactions in international maritime trade; the trade route stretching far beyond the city walls, into the depth of an unknown country—all of these elements of a trading city—proved to be crucial to British research into China's natural history. With minimal effort, British naturalists effectively reconfigured the entrepôt of Canton, turning it into a site of knowledge exchange and production. Urban practices in everyday life were transformed into powerful research tools and techniques.

Like tea, silk, and porcelain, plants, animals, and other scientific specimens from China arrived in London. However imperfect and incomplete these data were, they trickled in and marked the beginning of a new phase of British research into China's natural history. The British had contributed little to Western understanding of China's natural history until the late eighteenth century; but as the nineteenth century progressed, they would rival and perhaps outdo the French in this area. In the era of Old Canton, their primacy in the China trade afforded them access to China's plants and animals, and every aspect of the enterprise—the men, the ships, the networks, with all their opportunities and limitations—were inextricably linked to the China trade.

2

Art, Commerce, and Natural History

WESTERNERS in Old Canton came into contact with Chinese merchants, officials, servants, shopkeepers, and artisans from different social strata. Admittedly, the contact was often limited and controlled, yet it permitted the Westerners to carry on trade, maintain social connections, begin missionary work, and collect natural history specimens. The British in Canton took an interest in horticulture and natural history, both of which were respectable and fashionable intellectual pursuits in Britain. As residents in China, the British recognized the unique access afforded them to the natural riches of the vast empire, and they tried to maximize whatever contact they had with the Chinese to facilitate their collecting and research.

Confined in a narrow district, the British naturalists often had to rely on the Chinese to collect specimens, especially in the interior. Because the specimens thus procured were usually few in number or unlikely to survive the voyage to Europe, the naturalists had to find alternative or complementary methods of recording, preserving, and transmitting scientific data. Not surprisingly, verbal descriptions functioned as the most common and convenient method. In their letters or reports to the scientific community in Europe, British naturalists in China eagerly related the objects of nature they saw or observed; but these descriptions, often crude and imprecise, were hardly serviceable for technical research. It was not easy to describe in everyday language particular plants and animals in sufficient detail and precision that the naturalists' scientific correspondents could reproduce in their minds the objects in question without having seen them or any objects closely resembling them. Everyday language seemed to be more useful in communicating certain kinds of information, such as where and how a specimen was found or how best to grow a plant, than others, such as how to determine the identity of a plant.

This problem was, of course, not new. It had haunted natural historians for centuries; and by the end of the eighteenth century, highly standardized vocabulary had been developed to describe the observable characteristics of specimens. The precision of the vocabulary was such that educated practitioners could communicate effectively between themselves through written descriptions of specimens in technical vocabulary, for example, botanical Latin. This scientific language obviously was not for everyone; it required education, training, and practice. In the hands of qualified experts, according to many early nineteenth-century naturalists, written descriptions could be as reliable as, or even more reliable than, pictures. In their view, good verbal descriptions spoke directly to the rational mind, whereas pictures, especially the color ones found in popular texts, sacrificed knowledge for sensory pleasure.[1]

Yet others argued that pictures, with their strong appeal, could be employed to promote science and invite popular participation in natural history. In an unexpected way perhaps, the scientific activities of British naturalists in China supported this view. Unable or reluctant to describe specimens in technical vocabulary, the naturalists turned to visual representation as a useful alternative. Some British naturalists, prominently John Reeves, took advantage of the local industry of Chinese export painting and hired local artists to draw natural history specimens. The drawings were in a distinctive style and bore the visual signature of Chinese export painting.[2] This kind of collaboration was also common in eighteenth- and nineteenth-century India, where native artists, who had adapted their methods and style to the needs of their Western customers, were employed to depict natural history specimens.[3]

Natural history drawings thus produced invite us to consider two sets of issues. The relationship between art and science has generated much scholarly interest; and whether it is discussed in the context of museums, anatomical texts, or modern media such as film, the focus is usually on visual representation and translation in science.[4] In these discussions, natural history illustration features prominently.[5] The tradition of natural history illustration involved considerations of aesthetic effect and scientific authority. It therefore offers a fascinating and fruitful area for researchers investigating the historical relationships between art and science. That the works of many naturalist-illustrators were acclaimed both as works of art and scientific data only enriches the implications of the subject. The insects of Maria Sibylla Merian, the plants of Georg Dionysius Ehret, and the birds of John Audubon excited admiration from naturalists and were much sought after by art collectors.[6] The beautiful illustrations represented not only the talent of the artist-naturalists, but also the borderland where the cultural territories of science and art overlapped.

There is another important dimension to the natural history drawings under consideration. They constituted a site of cultural encounter, in which cross-cultural translation of ideas, aesthetics, and epistemology occurred. Modern historian Samuel Edgerton has used the odd discrepancies between seventeenth-century Chinese copies of European scientific illustrations and the originals to advance his controversial contention about the central role of a new visual culture in the Scientific Revolution. Bound by their visual conventions, the Chinese draftsmen in the printing establishment could not understand certain European techniques of pictorial representation, such as cutaways, and failed to produce intelligible copies of the originals. One might say that they couldn't *see* correctly—that is, "correctly" in a particular convention of realism.[7] Edgerton might well be right in his conclusion about the particular illustrations, but one must not overgeneralize his discovery to interpret all encounters of visual cultures.

Like most other cultural productions, visual cultures in art and science mixed, mingled, and hybridized. To be sure, artists and viewers often operated within certain conventions of visual culture, but they were also capable of crossing borders and inventing new forms and expressions. My emphasis is, therefore, radically different from Edgerton's. By analyzing the natural history drawings produced by British naturalists and Chinese artists, I wish to unravel the dynamic interplay of different visual cultures—Chinese and European, scientific and artistic—in natural history illustration.

Most of my examples are taken from John Reeves's collection of natural history drawings, which number more than one thousand and make up the largest collections of its kind. Before turning to the drawings themselves, however, I shall first introduce Reeves and his close associates in China and describe the social and cultural environment of the industry of Chinese export painting. Like many other businesses discussed in Chapter 1, export painting was a product of the expansion of maritime trade and the accompanying modern world economy of production, circulation, and consumption of goods.

John Reeves

During the times of Old Canton, the naturalist who did the most to increase British knowledge of China's natural history was John Reeves (1774–1856). He arrived in China in 1812 as a tea inspector for the Canton Factory of the East India Company, a lucrative position. Before his departure for China, he had been introduced to Joseph Banks through a relative and had received instructions for collecting curious plants and botanical lore. Unlike George T. Staunton, Reeves had had little scientific training. While a young man, he had entered a tea company to learn the trade and eventually made his way into the East India Company. When he joined the Canton Factory as Assistant Inspector of Teas, he was already in his late 30s, much older than a regular newcomer to the China trade. Many British and American traders came to China when they were only teenagers, hoping to stack up a fortune before their health was ruined by the reputedly injurious climate of the East. Reeves evidently had a gentle personality and was liked by his fellow merchants, though his name appeared only occasionally in their letters and memoirs. His own letters were usually short, but always cordial. Aside from his scientific pursuits, we know little about his life in Canton.[8]

In China, Reeves developed broad interests in science, including astronomy, but he devoted the majority of his attention to horticulture and natural history. Like most of his contemporaries, he practiced science as a gentlemanly pursuit, a respectable hobby rather than a true vocation. He did not dig deep into any of the sciences and never obtained more than modest familiarity with technical knowledge, such as taxonomy and anatomy, in natural history. He published only a few brief notes in obscure journals, collaborating with the missionary-sinologist Robert Morrison on a paper about Chinese *materia medica* and a report on a solar eclipse, neither of which received notice from the scientific community in Britain.[9] He also helped Morrison with the scientific terms in his English-Chinese dictionary.[10]

Reeves's scientific endeavors in China, however, earned him considerable respect among British natural historians at home. He was a member of both the Zoological Society and the Horticultural Society, and he was elected to the Royal Society and the Linnean Society in 1817. After his retirement from China in 1831, Reeves remained active in the scientific community in England as an expert on China's natural history. When the Royal Horticultural Society seized on the opportunity opened up by the Opium War to send a plant hunter to China, they sought advice from Reeves.[11] Back in China, Reeves's son John Russell Reeves, who had joined his father in Canton in 1827, continued the work of sending specimens to Britain, work for which he, too, would be elected to the Linnean Society and the Royal Society.[12]

John Reeves's success derived less from his learning than from his zeal, social connections, and economic resources. He purchased large numbers of new finds from the Fa-tee nurseries and sought rarities in the gardens of his Chinese friends, notably Puankequa II and "the Squire."[13] His interest in horticulture must have enhanced his image among the flower-loving Chinese merchants and strengthened his relations with them. His primary local associates, however, were his fellow British expatriates: John Livingstone, a surgeon to the Canton Factory, and Thomas Beale, a retired merchant, both living in Macao. Both men were among the most senior British residents in China when Reeves first arrived. Livingstone, a correspondent member of the Horticultural Society, occasionally reported to the Society his observations on Chinese gardening and other horticultural matters. As an experienced local expert, he provided Reeves with much help during the latter's first years in China.[14]

Reeves also benefited greatly from his friendship with the merchant prince and opium mogul Thomas Beale. Beale had lived in China since 1792 and was one of the most respected British residents in Macao. He was equally well known among the Chinese and Portuguese. Beale's life dramatically illustrates the roller-coaster course into which many an eastern merchant fell. While still a teenager, he came to China to join his brother's trade business. Through guts, luck, and shrewdness, he amassed enormous wealth, only to lose it and end up deeply in debt. Tormented and humiliated, he disappeared from his house one day in 1841. Weeks later, a few Portuguese boys playing on the beach were shocked to discover a decomposed body half buried in sand. It was old Thomas Beale. He had not seen England since his departure five decades before.[15]

In his active years, Beale was known for his hospitality. He welcomed any respectable visitors to his residence and usually struck them as a "somewhat formal" gentleman with "distinguished manners."[16] His mansion in Macao included a splendid garden of twenty-five hundred plants in pots, arranged in the Chinese fashion, and an even more famous aviary, a must-see for Western visitors to Macao.[17] The aviary, forty by twenty feet, contained hundreds of rare birds from China, Europe, Southeast Asia, and South America. Pheasants, magpies, parrots, and peacocks rivaled each other with brilliant plumage and spirited songs. George Vachell, the chaplain to the Canton Factory, described the sight in detail to his friend John Henslow, Professor of Botany at Cambridge University and mentor to Charles Darwin. There were about six hundred birds in the aviary at the time of Vachell's visit.[18] When the naturalist George Bennett stopped in Macao during his Pacific voyage, he was so impressed with Beale's garden and aviary that he devoted forty-five pages of his travelogue to describing their contents.[19] Many of the plants Reeves sent to England were from Beale's garden; as were the exotic birds he brought home with him, including the beautiful pheasant that would be named after him (Reeves).[20]

Collecting by Drawings

John Reeves's resources and social position rendered him a small-scale entrepreneur of natural history. Though most died en route, he shipped thousands of plants to Britain. He offered help and patronage to collectors sent out by the Horticultural Society of London, securing them access to the gardens of local Chinese merchants and introducing them to Beale and Livingstone. His most important contributions to natural history were, however, the botanical and zoological drawings that he sent to the Horticultural Society and his other scientific correspondents.[21]

Natural history illustration was an effective means of communication between a naturalist and his Chinese informants. A frustrated British visitor to Canton reported that "unless you know the Chinese name [in Chinese characters], or provide a sample of what you are in quest of, it is in vain making any enquiries." He added, "[A] painting of the plant would also work."[22] Indeed, drawings worked far better than any verbal description, including Chinese names, as a plant was often known by various vernacular Chinese names. "Collecting by drawings was the plan adopted by the collectors sent out by Mr. Slater," explained James Main, who visited China in 1792 to collect for Gilbert Slater of the East India Company. This method was supplemented by using "proper lists [of plants], with the English pronunciation of the Chinese names, and, which is better still, the Chinese character, if it can be had."[23] The plant names copied from Jesuit writings, however, were usually scholarly names; they were in widespread use only among the educated Chinese. Main found that Cantonese gardeners were familiar with few of the Chinese names he showed to them.[24]

The interest in Chinese flower drawings as scientific data was nothing new. In the 1690s, James Cunningham, a Company surgeon collecting specimens in China for Hans Sloane and James Petiver, assembled a collection of several hundred Chinese drawings of plants.[25] By the end of the eighteenth century, it had already become a common practice among British naturalists in China to gather Chinese drawings as scientific data or to employ Chinese draftsmen to illustrate botanical specimens.

John Bradby Blake, a supercargo at the Canton Factory, sent Banks a series of skillfully executed drawings by local artists, some labeled with their Chinese names.[26] In 1804, George T. Staunton wrote from Canton to the Court of the East India Company, "A Botanical Painter has been constantly employed in copying the plants, fruits and flowers of the Country, as they come successively in Season."[27] Indeed, when James Main arrived in Canton, the first thing he did was to visit the art studios and galleries in Canton. His purpose was to hire an artist to draw the plants he was to collect.[28] It was important to record the plants in this way. Most of the plants he collected would never survive the long voyage to Britain, and dried specimens had little horticultural value, as the color and shape of the flower were not preserved. Illustrations served as a convenient catalog of flowers; they facilitated further attempts to collect desirable plants. A similar method was adopted by Banks and Alexander Duncan. Banks owned a book of plant drawings by Chinese artists. To rank the desirability of the plants, he marked them with different numbers of crosses and then gave the book to Duncan, who collected specimens according to the suggested priority. The two British gardeners on the Macartney Embassy and, later, William Kerr were also equipped with the same handbook.[29] The book must have traveled back and forth between China and England several times.

Chinese Export Painting and the Hybridization of Visual Cultures

We can understand the practice of employing Chinese artisans to draw natural history drawings in the context of the China trade. With the expansion of the China trade, Canton had created an industry composed of artisans commissioned to produce exotic curiosities and commercial artworks—porcelain, furniture, paintings, miniature carvings, lacquered ware, and so on—catering to Western tastes.[30] The influence of European art on Chinese high painting in the Qing was limited. It did not spread much wider than the court artists who had contact with Jesuit artists in China, such as Giuseppe Castiglione.[31] Chinese export art, on the other hand, typically combined stylistic elements of both European and Chinese art. The style of export painting was a hybrid of Chinese genre painting and Western realism. The Chinese artists adopted (with various degrees of success) the techniques of Western realism, including perspective and chiaroscuro. Many also mastered the use of Western paints and tools, and became confident enough to execute large-scale oil canvases.

It is exceedingly difficult to trace the avenues through which the Western style was transmitted to Cantonese artists; written records are scarce. The establishment for finishing porcelain and for vase painting might have played an important role. It had been moved from Jingdezhen, the center for porcelain manufacture, about six hundred miles away, to Canton by the mid-eighteenth century, and numerous workshops were established in the city suburbs.[32] A British visitor to one of the workshops in 1769 reported that "in one long gallery we found upwards of an hundred persons at work in sketching or finishing the various ornaments upon each particular piece of the ware."[33] The workers ranged in age from old men to children of six or seven. These workshops often received specific designs—portraits, scenes, or coats of arms—from their Westerner clientele, a practice that can be traced back to the seventeenth-century Dutch traders in China, and the workmen finished the chinaware accordingly.[34] Samuel Shaw, who arrived with the first American ship to China in 1784, gave a Chinese artist the assignment of putting together the emblems of Minerva, Fame, and American Cincinatus.[35] The Declaration of Independence, portraits of George Washington, William Hogarth's drawings, and Christian art were among the subjects of porcelain paintings done in Canton.[36] The style of Chinese export painting on paper resembled the porcelain painting and developed in a similar way—that is, accommodating the taste of Western customers.

According to one account from the 1830s, there were about thirty studios of export painting in the neighborhood of the foreign Factories alone.[37] Indeed, Shaw already noted in the 1780s that "there are many painters in Canton."[38] Their products consisted mostly of portraits, local scenes, thematic series like the process of manufacturing porcelain, and plants and animals. These studios shared the same workshop culture with the other sectors of Chinese export art. The trades were usually family run, passed down from generation to generation. The studios adopted a guild system under which an apprentice studied under a master to learn the tricks of the trade, an education that often required years of practice. This thorough training sustained the acclaimed craftsmanship of Chinese export art. The chinaware painters employed, to quote a modern commentator, "assembly-line techniques to ensure uniformity."[39] A painting might pass through several hands before it was completed. One artisan traced the outline, another drew in the figures, a third man painted the background, and so on.[40] Similarly, the export painters used any techniques that fit their needs: copying, tracing, employing ready-made sketches of trees, houses, boats, or animals assembled in different ways to produce a different scene.[41] The finished paintings were shipped to Europe by the crateful, as well as carried home as souvenirs by Western visitors.[42]

Several practitioners were distinguished for their artistic accomplishments. Although none of these artists attained the fame of Rubens and the Renaissance masters, who ran their studios in a similar manner, some of them did impress their Western customers with the quality of their work. Lamqua, for example, competed with George Chinnery, a fine British artist in the academic tradition, for foreign customers in Canton in the 1830s. His brother Tingqua enjoyed similar success.

Lamqua even had his paintings displayed at the Royal Academy in London and several American exhibitions.[43] His studio in China Street, which employed eight to ten draftsmen, was always crowded with Western customers and onlookers. A customer needed to pay only fifteen or twenty Spanish dollars, or roughly £3, to have his portrait painted.[44] The studios of Tingqua and Sunqua, another major local painter, were nearby. A picture of Tingqua's studio shows three artists working diligently, one sitting after another, in a room neatly arranged and simply decorated.[45]

These artists left few traces in records other than their paintings. We know that Spoilum, the most important artist of export painting in the late eighteenth century, might have traveled to the West, but little additional information about him survives.[46] Tonqua, "the principal painter" in Canton at the time of John Reeves's arrival in China in 1812, left no written records apart from his answers to Reeves's bewildered questions about Chinese deities: How many figures of josses [i.e., deities] do you have in China? "[T]oo muchee, one two hundred."[47] What we can be certain of is that Reeves, or any other British naturalist, could easily find competent artists to draw natural history illustrations in the streets nearby the foreign Factories.

Picturing Nature

Reeves's natural history drawings covered a wide range of subjects: many insects; dozens and dozens of birds, shells, reptiles, and mammals; and hundreds of different kinds of plants, particularly ornamental varieties and fruits. His connection with the Horticultural Society explains the large number of fruits and garden plants; the Society was eager to discover and introduce Chinese plants of horticultural value, and Reeves's drawings served as catalogs of new finds as well as botanical data. The animal drawings were intended to be zoological data.[48]

John Reeves's special orders wouldn't have required his artists—Akut, Akam, Akew, Asung, and possibly some others—to alter their practice significantly.[49] Most of the local artists had practiced in the tradition of Chinese genre painting and were used to drawing the so-called bird-and-flower paintings and other subjects that required detailed depiction of animals and plants and careful use of color. Indeed, watercolors of animals and plants were among the most popular of Chinese export paintings. Europeans had been impressed with the beautifully rendered plants and animals since the eighteenth century; and no less an authority than Banks, himself an able botanical sketcher, would praise the scientific value of Chinese plant drawings: "the Plants painted by the Chinese, even in their Furniture, are so exact & so little exaggerated as to be intelligible to a Botanist."[50] George Bennett, the naturalist who wrote enthusiastically about Beale's aviary, thought that "the brilliancy of the Chinese colour for painting, &c. has often been very highly extolled as being superior to the European," and he expressed only admiration for the skills of the artists hired to depict Beale's animals and plants.[51] And years later, a British correspondent of the Zoological Society of London, referring to "the drawings of birds on Chinese fans, screens, &c.," concluded, "[They] are all more or less good representations of birds which exist in reality." "[W]ere [the Chinese artists] to pay more attention to minute detail, their drawings would give us a good idea of the ornithology of the country."[52]

Ironically, the style of Chinese export painting hardly appealed to the Qing literati, who preferred the more abstract *wenren hua*, the literati painting, and found the genre painting superficial and gaudy.[53] The paintings also differed greatly in style and subject matter from the illustrations in traditional Chinese herbals. Chinese herbals often included woodcut illustrations, but never colored plates.[54] The drawings were crude, and each picture usually dealt with only one subject (such as a plant, animal, or mineral) without any background. Unlike post-Linnean botanical drawings, a plant illustration in a Chinese herbal did not always focus on the flower. The illustration usually outlined the overall appearance of the plant, sometimes without showing the flowers at all. If the root of a plant was widely used as a medicament, the root would dominate the illustration. Some herbal authors did the illustrations themselves from observation; others reproduced the pictures from earlier books or probably left the task to the illustrators of the printing establishments.[55] Although Reeves knew no Chinese, he did know *Bencao gangmu* (1596), the most important of Chinese herbals, and must have seen its illustrations.[56] Because their quality was disappointing at best and because they included few plants found only in southern China, he could not have found them useful.

The illustrators for Chinese books worked for the printing houses, whereas the export painters practiced in the tradition of genre painting or decorative art. Even the best Chinese export paintings of plants and animals were usually set pieces, with insects, birds, and flowers poised in stylized postures in a conventional setting that might feature a rock, a tree peony, or a pond.[57] Although the carefully executed details gave the paintings an air of observational realism, and the artists' skill carried persuasive power, the subjects were nonetheless often imaginary. A plant might be decorated with flowers of striking colors, resembling no existing plant, all with a deceptive vividness. A butterfly might be a hybrid of the features of two or three different species or be an entirely fanciful invention.[58]

Thus, while Chinese artists possessed techniques highly adaptable to the kind of realism essential to natural history illustrations, they were not concerned with the "scientific accuracy" that an illustrator for, say, Curtis's *Botanical Magazine* was obliged to respect. They employed their skills simply to produce an aesthetic effect that would attract Western customers. To obtain the desired mode of representation, Reeves had to harness the Chinese artists' creative imagination in ways that were not incompatible with the established rules of natural history illustration. He had to keep it from transgressing the boundaries of "scientific realism." In addition, he also had to teach the Chinese artists what to pay attention to in depicting specimens. Evidence suggests that Reeves had the artists initially come to his house to execute botanical and zoological illustrations under his close supervision; and once the artists learned the principles, they worked at their studios.[59]

As a medium of scientific communication, natural history illustration often conveyed a descriptive immediacy and precision that no verbal description could achieve. Its "realistic" style invoked the authority of objective observation, of the visual sense.[60] However, natural history illustrations were not mirrors of nature, reflecting exactly whatever was put in front of them. They could function effectively as a medium of scientific communication because they relied on certain shared systems of encoding and decoding; they were interpretations of the subjects in a graphic lexicon. Like verbal descriptions, natural history illustrations also aimed to transcend the description of actual individual specimens; they were compositions of features deemed representative of the subjects. They thus stand for the idealized "general," "average," or "typical" specimens of the plants or animals.

Moreover, natural history drawings created for different purposes differed in their presentation and emphasis. Illustrations for botanical treatises emphasized the sexual organs of a plant that distinguished it from other species. Horticultural illustrations, on the other hand, highlighted the color and shape of the flower, because that was the feature horticulturists relied on to classify the breeds as well as the feature customers cared most about. Zoological illustrations of birds or mammals for the purpose of taxonomical classification were usually engravings of skeletons and particular bones. Those in general natural history books showed the fur/plumage or even the animal in its natural environment, as did William Jardine's plates.[61]

Without any training in natural history, the Chinese artists nevertheless managed to travel in the borderlands of different visual cultures—scientific and artistic, Chinese and Western. The practitioners of Chinese export art often worked on commission. They had long developed remarkable flexibility and the ability to meet the demands of different customers. Reeves's artists demonstrated a mastery of this craft. For Reeves's assignments, the artists had to depict actual specimens rather than copy existing pictures. Their experience in portraiture and insect drawings must have been particularly helpful. They worked individually instead of as a team, because the assembly line or other mechanical methods obviously would not have worked in this case.

Natural History Illustration As a Site of Cultural Encounter

Unlike regular export paintings, few of Reeves's drawings had a background. The plant drawings were done in the tradition of horticultural illustrations, admiringly emphasizing the whole flowers and covering different varieties of the same plant. Some of them laid out the floral details needed for classification (in other words, generic characters), which was of course a new thing to the Chinese artists.[62] The pictures of birds, mammals, reptiles, and crustaceans resembled the plates in multivolume natural history books in their emphasis on the general appearance rather than the anatomical details of the subjects, though some of them included close-ups of body parts (such as the mouth and teeth, or the foot).[63] Reeves's knowledge of natural history determined to a great extent the scientific value of the drawings. He seems to have possessed little comparative anatomy or other technical means for identifying zoological specimens, so some of his animal drawings appealed more to the eye than to the mind.[64] His Chinese draftsmen carried out his instructions with remarkable skill and observational power,

Art, Commerce, and Natural History   53

but they lacked the knowledge to contribute to Reeves's scientific judgments.

The drawing of the betel palm demonstrates the techniques of representation and how a large amount of information could be condensed in a scientific illustration.

On the left side of the drawing stands a mature palm tree that bears two bunches of fruits. One bunch of fruits has ripened and turned orange-brown; the other is still green. This image not only shows the appearance of a "typical-looking" betel palm—a slender single-trunk palm with leaves borne at stem apex—but also gives information about how and where the fruits grow and the comparative appearances of ripened and immature fruits. In all likelihood, the illustration was not an exact copy of a real tree that happened to have two bunches of fruit at different stages of ripeness. Instead, it was a composition of different elements chosen to convey scientific information.

Now, look at the lower half of the drawing. It consists of two groups of images. On the left is an enlarged image of a cluster of oval-shaped fruits on a freshly broken branch, the tip of which points slightly downward. The fruits closer to the base of the branch seem larger and riper than their successive neighbors to the right. The comparison is skillfully highlighted with different shades of brown, orange-yellow, and green. A twig, longer than any other, protrudes out of the upper-middle part of the dense cluster of fruits. It bears many tiny buds and flowers. Thus, the branch as a whole illustrates the pattern of distribution and the different stages of maturity of the flowers and fruits.

The scientific purpose of this drawing becomes unmistakably clear when one looks to the lower-right corner of the drawing, where one finds a set of small images of a dissected seed-bearing fruit and the anatomic parts of a flower. The anatomic details are carefully laid out to assist botanical identification. Leaving the details behind, move up again to the huge blade of leaf that cuts across the drawing diagonally. The leaf is folded over, with a section of its backside visible. It looks like a real, flattened specimen mounted on a sheet of paper.

The subject matter, composition, and purpose of the drawing are far different from those of regular export painting, and the Chinese artists must have regarded them as somewhat peculiar. The palm tree was hardly a popular subject in Chinese painting, not even in the south. The scales of the different images in the drawing vary so enormously (a tree appears much smaller than a leaf) that the drawing cannot pretend to be a coherent representation of a scene. A tree, a leaf, and a bunch of fruits are juxtaposed without any specification of their actual physical relationships. Then there are the dissected fruit, seed, and flowers. These features enable the drawing to function as a scientific illustration. Under Reeves's instruction, the Chinese artists did successfully translate physical objects into scientific illustrations. The process required not only technical competence in drawing but also an imaginative selection and combination of characteristics of the plant. Like this example, many of Reeves's drawings succeeded well as scientific illustrations.

Despite the apparent conformity to the idiom of natural history illustration, however, incidental details of Reeves's drawings betray their origin. The style of Chinese export painting is registered in the bold but skillful use of color, the solid, even stiff outline, the dexterous brushwork, and the flat appearance resulting from clumsily applied shade and light.[65]

The Chinese style speaks out loudly when the subjects are the favorites of traditional Chinese painting, such as the plum tree and the crane. In these cases, conventional mannerism manifests itself. With its crooked branches, exaggerated knots, and cracked bark, the plum tree exhibits the classic aesthetic devices of Chinese flower painting.[66] Similarly, the gesture of the red-crowned crane—standing upright with one leg hidden beneath the body, its neck gracefully curved, its eyes gazing peacefully ahead—comes straight out of the Chinese tradition.[67] The plum tree and the crane were symbols of endurance and longevity, respectively, in Chinese culture. They had long-acquired typical modes of representation, just as the lion invariably appeared regal, and the swan elegant, in Western natural history illustrations.

Visual Authority and the Production of Knowledge

The Chinese artisans probably saw Reeves as an eccentric customer. They certainly did not understand the scientific value of their own work. To them, painting was first of all a trade; and whatever artistic merits they assigned to the drawings (perhaps not very much), they understood them to be commodities as well. British naturalists saw the drawings in a different light. Regarding the animals and plants of China, Reeves's drawings were often the only source of information—

in many cases, there were no specimens, no verbal descriptions of any kind, only the pictures.

Reeves's drawings, therefore, were not natural history illustrations in the usual sense. However impressive the drawings of James Sowerby, John Gould, and W. H. Fitch were, they did not stand on their own, but worked in tandem with texts and had specimens to fall back on.[68] They were meant to complement verbal descriptions, although, as powerful images in their own right, they went on to have a life of their own in art as well as in science. Reeves's drawings, on the other hand, *were* scientific data and often the only scientific data the Europeans had of certain Chinese animals and plants. There was little possibility of identifying specimens on the spot in China. Reeves's scientific learning did not allow him to make informed conclusions from the specimens, a situation only worsened by having no access to any herbarium, museum, or science library.[69] Consequently, Reeves could provide his colleagues in Europe with little additional information.

Joseph Sabine, John Lindley, and other British naturalists often had to work exclusively from the pictures, without consulting specimens. Empirical confirmation of some of Reeves's drawings (as when a plant did survive the crossing) certainly played a role in winning confidence for those without supporting specimens. Sabine remarked that "[of] the correctness of the representations in these drawings I have little doubt, because the resemblances of other plants figured by the same artist, which have blossomed here, are perfect."[70]

Nevertheless, it is still significant that the products of Chinese artisans and a naturalist-collector with limited botanical expertise actually attained a visual authority recognized by the scientific community in Europe. Sabine himself well knew that many of the chrysanthemums transplanted from China bore flowers that looked very different from those represented in Reeves's drawings with the same names. He attributed the cause of this difference not to the possible defect of the Chinese artist's skill or observational power, nor to Reeves's mistake or want of knowledge, but to the different ways the Chinese and English gardeners treated the flowers and no doubt English soil, sun, and climate.[71] Sabine gave Reeves and his Chinese artists a great deal of credibility indeed. As late as the nineteenth century, the network of scientific information in natural history remained fairly open and flexible.[72]

Reeves's celebrated series of fish drawings, now in the Natural History Museum of London, testifies to the achievement of his collaboration with the Chinese draftsmen.

Fish were notoriously difficult to preserve; they decayed and lost color quickly. Neither dried nor pickled fish revealed much of their original appearance. In fact, they often rotted to a state of disintegration during a voyage of four months in warm weather or were too big for regular bottles of spirits.[73] Not surprisingly, the Europeans knew little about the fish in Chinese and Japanese waters until trained naturalists, including Philipp Franz von Siebold, collected specimens in the region in the second quarter of the nineteenth century. Linnaeus, for example, knew only nine species of Chinese fish. John Reeves's fish drawings, which amounted to approximately 340 species, mostly from the Canton and Macao markets, constituted an important source of information.[74] They were watercolors executed with superb craftsmanship, most of them drawn from living or freshly killed specimens.

To reproduce the iridescent rainbow hues of fish scales, the artists made use of silver and gold powder. They worked on the paintings in their studios at a high speed and brought the finished work to Reeves for inspection.[75] In reference to the drawings, Sir John Richardson, a distinguished ichthyologist, reported to the British Association for the Advancement of Science in 1845: "These drawings were executed with a correctness and finish which will be sought for in vain in the older works of ichthyology, and which are not surpassed in the plates of any large European work of the present day. The unrivalled brilliancy and effect of the colouring, and correctness of profile, render them excellent portraits of the fish they are intended to represent."[76] Indeed, even "further details of a technical kind," as long as they were "sufficient[ly] large to be conspicuous to the naked eye," the "Chinese artist has seldom failed to represent them."[77] Richardson named eighty-three new species based solely on the drawings.

To be sure, Reeves's drawings were not furnished with complete details for precise classification, which required descriptions of the dentition, the number of the gill-rays, and other minor characteristics, so Richardson was taking some risk in making the identifications. Due to the paucity of data, however, drawings by native artists had always been important to European knowledge about fish of East Asia.[78] In fact, much of what British naturalists knew about the fish of India had come from Thomas Hardwicke's fish drawings by native artists.[79]

Therefore, Richardson had precedents. His paper was a major ichthyological contribution, listing more than six hundred Chinese and Japanese species, which was substantially more than in any previous list of the same kind.[80]

John Reeves's practice was not singular. As we have seen, John B. Blake, James Main, and William Kerr, among others, engaged Chinese draftsmen in Canton to draw botanical illustrations as catalogs of Chinese ornamental plants.[81] Thomas Beale also had the best of his birds and flowers faithfully recorded in paintings.[82] About the same time, Sir Thomas Stamford Raffles in Singapore hired a Macao Chinese and several local talents, who were probably overseas Chinese, to depict his collection of animals and plants.[83] In the mid-nineteenth century, Theodore Cantor, Henry Hance, and Robert Fortune all commissioned Chinese artists to draw natural history illustrations.[84] Apart from their scientific value, these drawings were also celebrated for their artistic merits. The mixing of the Western realistic tradition and the genre of Chinese export painting produced distinctive styles that were at once precise and attractive.

REEVES'S DRAWINGS constituted an important source of British knowledge of Chinese plants and animals in the early nineteenth century. They provided British naturalists with unique scientific data. British naturalists in China employed Chinese artisans to facilitate their research; the Chinese artisans, who had already absorbed elements of Western realism, quickly adapted their skills to the requirements of their new customers. Both sides were actively exploiting the opportunities brought about by the transaction, for their respective purposes. Both sides were cultural agents generating and circulating hybrid cultural productions. In Reeves's drawings, art, commerce, and natural history converged. And they should be seen as a synecdoche for the broad cultural contact that involved the exchange of goods and currency, the accommodation of tastes and ideas, the extension of human relations, and the encounter between empires. The tradition of visual representation in natural history allowed British naturalists in China and in Europe to communicate scientific information, even when other means proved unsuccessful. But it was the Chinese artisans who made the process possible.

II
THE LAND

3

Science and Informal Empire

AFTER ITS DEFEAT in the Opium War, China was forced to open treaty ports for trade with the Westerners. The British were no longer confined to Canton; they occupied Hong Kong and gained access to five coastal ports in 1842. Driven by trade and diplomatic interests, the British, assisted by the French, again bullied Qing China, which was already plagued with natural calamities, fiscal difficulties, and a succession of uprisings. In northern China, the Yellow River repeatedly flooded and left millions of people homeless.[1] The Taiping Rebellion (1850–1864) overran much of southern and central China and would eventually claim twenty million lives. In 1856, the Arrow War broke out between the British and the Qing government. After a series of wars (collectively called the Second Opium War), the Anglo-French armies occupied Beijing, looting and torching the summer palace of Yuan Ming Yuan in 1860.

Desperate for a break, the Qing government signed unequal treaties that granted the Westerners generous legal and diplomatic rights. Westerners were now allowed to set up commercial and diplomatic posts in inland China. The number of the treaty ports would gradually increase to about forty by 1911, scattered across China proper and Manchuria. Most of them were located along the China coast and the Yangzi River—the areas known as the British sphere of influence.[2] Not only did the British now have more freedom to travel about the treaty ports and, as the century progressed, in the interior, but they also successively established diplomatic and missionary institutions in many parts of China. New opportunities attracted thousands of merchants, diplomats, and missionaries from Europe and the United States.[3]

British research into the natural history of China experienced marked changes triggered by these political upheavals. Greater access to different parts of China and more intimate contact with the Chinese opened up new areas of research, one of the most prominent being economic botany. In the late eighteenth and early nineteenth centuries, Joseph Banks and his China correspondents had been systematically collecting information about the plants from which commercial products were made, but horticulture and ornamental plants were the subjects of most of their research. The British quest for garden plants in China never diminished. After the middle of the nineteenth century, however, their horticultural focus gradually switched from Chinese gardens and nurseries, whose treasures seemed to have been exhausted, to wild plants. Robert Fortune, who came to China for plant hunting in the 1840s and 1850s, was perhaps the last to make most of his valuable discoveries in the gardens and nurseries in the cities. By contrast, his successors at the turn of the century, such as Ernest Henry Wilson and George Forrest, went directly into the mountains of Hubei, Sichuan, and Yunnan.[4] Meanwhile, increasing contact with the Chinese drew the naturalists' attention to a wide variety of plant products used by the natives of which they had been unaware. The thought that these investigations might prove to be of great commercial importance added significance to the naturalists' research.

Scholars have not neglected the connections between imperialism and economic botany. It is widely accepted that the British botanical empire, with a nerve system made up of Kew Gardens and the colonial gardens, expanded its tentacles deep into all possible territories in search of plant products of economic value for the empire.[5] While this description remains partly valid, the peculiar historical context of the British in China forces us to modify this view. One problem is that extant literature on scientific imperialism has dealt primarily with the colonial context. In the China region, however, the only British colony was the tiny island of Hong Kong and its immediate vicinity.[6] Despite considerable internal and external pressures, the Qing government remained largely autonomous. Here the concept of "informal empire" may serve as a heuristic guide to describe British scientific imperialism in China. The concept originally refers to the economic control, often in the name of free trade, that an imperial power exercised beyond its formal territories. For our purpose, we shall not focus so exclusively on trade or the economic aspect of imperialism—though they remain important components of our historical interpretation—but shall venture into less explored topics, such as informal empire and its relations to the networks of scientists, the political economy of knowledge, and the dynamics of cultural contact.

The foreign imperial powers, particularly Britain, extorted from the Qing government extraterritoriality, concession leases, and other unequal treaties by persuasion, pressure, and gunboat diplomacy. British presence in China owed much to its imperial domination, and that historical condition helped shape the aims, means, and possibilities for scientific research. The diplomatic and other institutions whose missions were not primarily scientific proved to be instrumental in British scientific imperialism in China. The major establishments that took part in the natural historical research were the British Consular Service in China, the Chinese Imperial Maritime Customs, the Protestant missionaries, and the Hong Kong Botanic Gardens.[7] The British Consular Service in China was an official diplomatic institution. Its consulates could be found in most of the treaty ports, and together they formed the largest consular network in the world.[8] The Chinese Maritime Customs was a department of the Chinese government, but it fell under British influence. Its institutional structure was even more extensive than the British Consular Service in China. Only by bringing these nonscientific institutions to the center of the historiographic stage can we understand the scale, purpose, and practice of the British botanical empire in China.

The naturalists, who were widely separated in different provinces, needed networks for gathering, analyzing, and distributing information and specimens among themselves. This problem of the naturalists reflected a general concern of Westerners in China. Whether they were missionaries, merchants, or diplomats, they all found themselves involved in collecting and circulating an ever-increasing amount of information about the "mysterious" empire and its "singular" people. Relevant, accurate, and sufficient information about China seemed vital for many reasons, and the result was the emergence of institutions that made the collection and diffusion of data a principal objective. All of this activity worked together with the efforts to "unveil" China, which was motivated by practical considerations as well as by an orientalist desire to explore the vast Middle Kingdom and to produce "objective" factual knowledge about it.[9]

The intensification of British imperialism in the China region that occurred in the wake of the first Opium War added another dimension to research in China. Collecting and possessing "useful" information about China laid the groundwork for British imperial surveillance and the exploitation of economic opportunities. The British Consular Service in China, so enormous as to outstrip immediate relevance for trade, was not merely "an expensive luxury," as has recently been suggested.[10] It was frequently mobilized for the accumulation of intelligence or "facts"; it served a critical function in British informal empire. The research into China's natural history epitomized the characteristics of British research on China in general: it aimed at producing "factual" and "useful" knowledge about China—in this case, its natural products. Consequently, the naturalists were able to use the nonscientific institutions they belonged to for natural historical research.

However, there were real constraints on British informal empire in China, especially beyond the coast, the treaty ports, and the Yangzi River. Any study of British scientific imperialism in China must acknowledge this political reality and its implications for scientific research. The naturalists frequently found themselves lacking the ability to coerce the Chinese into providing information; in such cases, the mode of collecting information was one of negotiation, which did not always go in favor of the naturalists. The fact was that the British Empire's advantage in global power politics did not permeate locally in China or indeed in many colonies. It is therefore necessary to examine the tactics, meanings, and power relations involved in these negotiations because they set the framework within which the British collected information in China. The ideology and practice of collecting information about China, as we shall see, were interconnected with scientific imperialism and the naturalists' belief in the cognitive superiority of the modern Western tradition of factual knowledge.

Hong Kong Botanic Gardens

As the British Empire grew, Kew Gardens and its satellite gardens in the colonies formed a network that circulated living plants, specimens, and information across the globe. They also participated directly in many imperial and colonial enterprises that required the development of plantations of crops of economic value, such as tea, sisal, and cinchona trees, in the colonies.[11] In the China area, the British had only one colonial botanical garden, the Hong Kong Botanic Gardens, which maintained close ties with Kew Gardens and worked toward goals similar to those of other colonial gardens. For various reasons, it never became the central institution of British research into the flora or the economic botany of China. Because colonial gardens had been crucial to the building of the British botanical empire in other areas, the disappointing accomplishment of the Hong Kong Botanic Gardens needs further explanation. The example will also usefully remind us of the tensions, conflicts, and competitions that different imperial institutions, despite their intended cooperation, experienced with each other. The British Empire was not a coherent, smooth-running whole.

In its early colonial years, Hong Kong gained notoriety for its devastating tropical climate, which was thought to be responsible for the high disease and death rate among the British troops stationed there. "During July August & September [1850], we buried about 300 men," a soldier lamented.[12] That took care of a quarter of the total number of the troops. In hopes of improving the climate, the colonial government of Hong Kong decided to forest the island by planting large numbers of shade trees, a practice common on many tropical colonial islands.[13] The Hong Kong government also wanted to establish a pleasant garden retreat for the British residents. As a new colony, Hong Kong had no fine gardens to boast of, and its British residents could only admire the gardens of St. Helena, Mauritius, Calcutta, and Ceylon. Through the Colonial Office in London, the Hong Kong government sought help from Joseph Hooker, Director of Kew Gardens, by asking him to recommend a good gardener who could take the responsibility of developing and superintending the projected garden. Charles Ford, a practicing gardener who had served at several private and public gardens in Britain, was selected to fill the position. He arrived in Hong Kong in 1871.[14]

In spite of this promising start, Ford's career soon suffered the same difficulties in which many superintendents of colonial gardens found themselves, sandwiched between colonial governments and Kew Gardens. Ford's responsibilities included not only developing the Hong Kong Botanic Gardens but also the afforestation of the island. Indeed, on his retirement, when Ford looked back at his career of thirty-one years in Hong Kong, he regarded foresting the island, instead of developing the Gardens, as his "magnum opus."[15] To the Hong Kong government, Ford's first duty in managing the Gardens was to create and maintain a beautiful public garden for Westerners in Hong Kong. It would be an exaggeration to compare Ford to someone like Ferdinand von Mueller, a noted botanist and Director of the Melbourne Botanic Gardens, whose enthusiasm for scientific research led him into serious conflicts with local dignitaries who cared more about colorful flowers than dried specimens.[16] Nevertheless, Ford, like Mueller, frequently complained to Hooker about his difficult situation. So much of his time was spent on planting trees and growing flowers that botanical research became a rare luxury. A particularly unsympathetic governor was said to care little about the Gardens other than demanding a regular supply of vegetables for the dinner table of "his tremendous family."[17]

The Hong Kong government was not simply being backward, myopic, or vulgar—although that seems to have been the assessment of Ford and his scientific friends.[18] The administration was convinced that no botanical research, even if related to economic botany, would profit the colony.[19] It saw no possibility or reason to develop plantations of any kind on a minute mountainous island whose economic prospects rested entirely on trade, shipping, and commerce. It seemed clear that any fruits of botanical research sponsored by Hong Kong would always go to some other British colonies. To the Hong Kong government, the Gardens was simply a "public recreation ground" where Western residents could stroll after a day's work and attend concerts in summer.[20] It thus had neither a scientific nor an economic function.

The view taken by the Hong Kong government on its botanical garden contradicted that of Kew Gardens. Because the Botanic Gardens officially fell under the administration of the Hong Kong government, Kew Gardens had only limited control over its management. In order to exert an influence, Joseph Hooker had to pull strings through his connections to the Colonial Office in London, which supervised all the colonial governments, and through his patronage of Ford. Hooker had a clear vision of what the colonial Gardens should be like, and he expounded the idea in a letter to the Colonial Office.[21] Hong Kong, according to Hooker, occupied a unique position in discovering "the economic and scientifically interesting vegetable productions of China."[22]

The notion that the Gardens was simply a pleasure ground ran against "[Hooker's] idea of a Botanical Garden in any sense."[23]

On the contrary, the Hong Kong government, he argued, should allow Ford to make occasional expeditions and encourage him to transform the Gardens into "the channel of communication for all matters of botanical interest connected with China."[24] In a separate letter, Hooker pointed out to Ford that little was known about the "indigenous vegetation of South East Asia and of the useful application of its plants."[25] He instructed Ford to set up a herbarium and to visit the accessible places in China "for the purpose of collecting plants and of obtaining information about the sources of the drugs, woods, &c." that were used by the Chinese.[26] Hooker further noted that botanical research of this kind was beyond the ability of any individual or institute, and he suggested that Ford "enlist the co-operation of traders, merchant captains, sea surgeons and others" in his inquiries.[27] In unequivocal tone, Hooker reminded Ford to make the Botanic Gardens the headquarters of the investigation into the flora of China, "whether scientific or simply *utilitarian.*"[28]

With these instructions from London, Ford's official responsibilities now included exploration and scientific correspondence, besides gardening and afforestation.[29] He was able to set up a modest herbarium, together with a small library of botanical and gardening books.[30] Hooker's goal, however, was never fully realized. In spite of Hooker's pleadings, Ford still had little time and few opportunities for botanical research.[31] Ford's somewhat passive personality ("a quiet modest young man," according to a friendly observer) also prevented him from confronting the Hong Kong government, whose support for scientific research remained more nominal than material.[32]

Another reason why Ford and his establishment did not become the hub of botanical correspondence in China was that one already existed. Henry Hance of the Consular Service had been studying the flora of China since the 1840s, and he had long established a vast personal herbarium, a wide network of botanical information, and a well-deserved reputation as an authority on Chinese plants. Indeed, Ford learned his basics of botany—identifying plants, preserving specimens, and so on—from Hance and constantly consulted him on botanical matters.[33] If anything, Ford was incorporated into Hance's already extensive botanical network rather than the other way around. Even when Hance died in 1886, Ford was unable to fill the gap. How did a consular officer at an insignificant post acquire so much botanical authority and resources? Hance's scientific career illustrates the scale, influence, and effectiveness of the network of scientific data forged by Western residents in China and leads us to consider some of the salient characteristics of British scientific imperialism in general.

Henry Fletcher Hance

Except for brief stints in Amoy and Canton, Henry Fletcher Hance (1827–1886), Vice-Consul at Whampoa, was trapped in that minor post and had no chance of promotion because he knew no Chinese.[34] Early in life he went to schools in London and Belgium and mastered Latin, Greek, French, and German. As if that were not enough, he taught himself botany in his spare time. He came to China with his father in 1844, entered the Civil Service a few years later, and returned to England over 1851–1852 for a visit.[35] By then he had begun publishing botanical papers and had received an honorary Ph.D. from the University of Giessen in Germany. During his stay in England, he met William Hooker, Director of Kew Gardens, and would become William and Joseph Hooker's chief correspondent in China. On New Year's Eve of 1851, Hance proposed to his fiancée, "the playmate of my childhood," by a lake in Kew Gardens and subsequently brought her to China.[36] They soon had a large family. Financially stressed, Hance never managed to visit England again. He died thirty years later, a world authority on Chinese plants.[37]

A taxonomist, Hance never did any notable fieldwork, except during a sojourn in Amoy in the autumn of 1857.[38] The location of his post at Whampoa was probably the worst fieldwork site one could have in China, for the whole Canton area was denuded of wild vegetation and the place had been worked thoroughly.[39] Yet Hance realized that the flora of much of China remained unexplored and that collecting specimens and information was critical to research. Driven by his zeal for botany, Hance soon built an extensive network of scientific correspondents across China, with his colleagues in the Consular Service as his core associates. Although Hance was eccentric, forthright, and probably intemperate (qualities that must have hurt his professional career), he was also kind, generous, and principled.[40] He read omnivorously and owned a well-stocked personal library.[41] The critical period of his networking in China was probably in the late 1840s and early 1850s when new members of the Consular Service still often stopped in Hong Kong and Canton for a few months to learn the Chinese language while waiting to be assigned to their posts. Hance impressed them with his learning and kindled their interest in collecting botanical specimens, if not necessarily in studying botany itself. As his enthusiasm for botany became widely known, his network of scientific data in China expanded very much on its own.[42]

British naturalists in China respected Hance greatly. Robert Swinhoe, the foremost British zoologist in China, and Hance were consular colleagues, friends, and fellow naturalists.[43] William Hancock, botanist and a Chinese Maritime Customs commissioner, told Joseph Hooker that "[my] personal regard for Dr Hance, (independent of my admiration for him as a botanist), is such, that I preserve with care all his letters."[44] When Theophilus Sampson, Hance's botanical sidekick, traveled extensively in the province of Guangdong as an agent for the infamous coolie trade, he also collected for Hance.[45] Edward H. Parker, a consular colleague in Canton, studied botany with Hance and never lost his interest in the subject.[46] Hance influenced deeply these and many other people who came in contact with him. His dedication to botany earned him wide respect even among non-British Westerners in China, and they, too, corresponded with him and sent him specimens. Encouraged by him, Francis B. Forbes, an American merchant, launched a project of cataloging all Chinese plants then known to science. The project, soon taken over by the British, would culminate in the authoritative series of *Index Florae Sinensis* (1886–1905).[47] Emil Bretschneider, Physician to the Russian Legation in Beijing and a man of wide learning, became a good friend of Hance's on the basis of their common interest in botany and respect for each other's scholarship.[48]

Thus Hance received plants from his correspondents all over China and built up a splendid personal herbarium comparable to the major research institutes in Europe in its collection of Chinese plants, which included numerous type specimens.[49] His wife assiduously arranged and labeled the specimens.[50] Hance relied little on research centers in Europe for Chinese plants, for his personal network of correspondents in China brought him numerous specimens and gave him better access to, and claim for, authority on the plants in south and east China than did any botanists in Europe. Indeed, he was a major source of Chinese plant specimens for research institutes in Europe, using the many duplicates received from his network to exchange with them specimens for taxonomic comparison. He published more than two hundred papers on Chinese plants and related subjects, most in major journals. Hance's network of correspondence was global, reaching Asa Gray in the United States, Carl Maximowicz in St. Petersburg, and many major botanists in Belgium, Germany, France, and the European colonies. He was particularly closely connected to Kew Gardens, sending them the best duplicates of his Chinese plants, and he regarded Joseph Hooker and George Bentham as his principal botanical friends in Britain, who reciprocated this confidence with efforts to obtain promotion for him as well as routine scientific assistance.[51]

The relationship between Hance and the botanists in London wasn't always smooth, however. Some of the conflicts arose from his earnest personality. Once convinced that he was right, Hance did not hesitate to challenge the authority of the great and powerful. Consulted by William Hooker in 1862, Hance reasoned forcefully with the grand old man about the latter's plan and execution of sending collectors to East Asia, trying to convince Hooker that he had picked the wrong man on a previous attempt and that his design and management of the current plan were flawed.[52] Hance's confidence derived from his familiarity with the flora and situation of China, a confidence shared by Bretschneider and other competent naturalists in China. Bretschneider once became annoyed by William Thiselton-Dyer, then Director of Kew Gardens, for doubting his (and Hance's) authority on the identity of a Chinese plant of economic value.[53]

During a dispute over specimens a few years earlier, Thiselton-Dyer faulted Ford for sharing a new collection of plants with Hance before sending them to Kew; he also accused Hance of infringing the property right of Kew thereby and of trying to secure a monopoly on the botany of China.[54] All scientific institutes carefully guarded their property and intellectual rights. They wanted to own good specimens of new plants, especially type specimens, and to publish reports on the novelties. Hance and Kew, however, had previously reached an agreement that to take advantage of the freshness of the specimens he could describe Ford's new plants. Convinced of his own innocence, Hance vigorously defended himself and indeed reprimanded Thiselton-Dyer as though the Assistant Director of Kew was merely an erring junior colleague.[55]

The tension between botanists in the metropole and in the colony could also involve disagreements over scientific theory and practice. These were accentuated by Bentham and Hooker's attempts since the 1860s to stabilize the wildly diverse plant taxonomy and nomenclature. Not only did they drastically delimit species and genera in their *Genera plantarum* (1862–1883), but they tried to consolidate control over plant taxonomy by adopting a number of measures that favored major science journals and large herbaria in admitting new species. This move tipped the balance of power toward major scientific institutions in the metropole and put provincial and colonial botanists in a structurally subordinate position.[56] Although Hance's authority on the flora of China was hardly threatened by this new development in systematics, he, like many others, found it taxonomically impractical. His view was further compounded by his ambivalent attitude toward Darwinism.

The storm over Darwinism in Britain stirred few ripples in the British community in China. No heated debates broke out.[57] British missionaries in China continued holding on to their belief in natural theology, and Darwin was rarely mentioned in their publications throughout the rest of the nineteenth century. The best British naturalists in China, on the other hand, immediately recognized the power of Darwin's theory of evolution. Swinhoe, a regular correspondent of Darwin, readily embraced Darwinism, just like many of his ornithologist colleagues at home. In fact, Swinhoe was among the ninety colleagues and friends to whom Darwin presented a copy of the first edition of *On the Origin of Species*.[58] Hance's mind was of a more pensive cast, and his attitude toward Darwinism was more complex. Although he considered himself a fellow traveler of the Darwinians, he worried about the extreme tendency of combining species based on abstract notions of variation and affinity. This practice was so arbitrary, overly theoretical, and deficient in empirical justification that it could render itself vulnerable to the adversaries of Darwinism. They could argue that the "*hypothesis* [was] incapable of proof" and would, "by its nature, never become a recognized *fact* in science." Therefore, Hance urged that observable distinctions between plants should not be dismissed on account of a theory.[59]

Despite its appeal to observation and empirical science, Hance's comment actually originated from a deeper concern than mere plant taxonomy. In private correspondence, he and Hooker had a "candid" exchange over views in connection to Hooker's presidential address to the British Association for the Advancement of Science in 1868. While praising Hooker's paper as a masterly defense of Darwinism, Hance drew a line between himself and the more committed Darwinians. He considered Darwin's theory of evolution "unassailable . . . within certain limits," but refused to follow Hooker and Darwin all the way in their recent opinions. In his view, when the theory was pushed too far, there was only one logical end—materialism. Hance was disturbed by the looming specter of "Carl Vogt and his school."[60] In response to Hooker's confession that he now held the idea of the afterlife more as a hope than a faith, Hance explained that he had found home for twenty years in the liberal theology of David Friedrich Strauss and other German biblical scholars. Intellectual disagreements notwithstanding, Hance's friendship with Bentham and Hooker remained warm until his death.

The scientific career of Hance demonstrates the possibility for naturalists to develop powerful and extensive scientific networks on the institutional base of nonscientific organizations in China. Hance had no resources other than his energies and personal qualities, by which alone he mobilized his friends and colleagues in the Consular Service and other organizations to collect scientific data for him. But his success also depended on the particular nature of the social world of British residents in China and of the diplomatic and other organizations. It was a small world despite its geographical range; information and influence could be transmitted quickly and broadly. Everyone knew that Dr. Hance welcomed specimens and that he never refused to offer help with botanical matters. Specimens poured in from all directions to Hance's house in the tiny town, where he, a frustrated consular officer fond of drinking, reading, and children, carefully identified the plants and introduced them into the world of botany.

Institutions of Informal Empire

Although the cohort of Hance's scientific correspondents in China consisted of his colleagues in the British Consular Service, some other institutions in China—the Chinese Maritime Customs, the Protestant missionary organizations, and even the merchant class—also provided talent and infrastructure for natural historical research. The two government services employed large numbers of British officers. Between the 1840s and the 1890s, the British Consular Service in China employed more than two hundred officers, excluding lesser employees and the many Chinese writers. By 1880, they had opened consular establishments in more than twenty cities.[61] The other political institution, the Chinese Customs, was actually a government agency of China, founded in 1854 to run the maritime customs at the treaty ports.[62] Under British pressure, the actual management of the organization went into British hands from the very beginning. Its officers, from the inspector-general down, were mostly British, but also included other Europeans and Americans. In 1896, the Customs had 679 Western employees, of whom 374 were British, 83 German, 51 American, and so on.[63]

Both services adopted similar methods of recruiting new members after about 1860. They selected well-educated British and Irish young men (and in the case of the Chinese Customs, continental Europeans and Americans as well) through examinations.[64] Recruits from Scotland and Protestant Ireland were noticeably overrepresented, reflecting the general profile of British subjects in the imperial services. The Chinese Customs Service drew heavily on Protestants from Ireland, because its inspector-general, Robert Hart, came from that background. Scottish universities, of course, supplied many of the medical personnel. Patrick Manson, the so-called father of tropical medicine, was a graduate of the medical school at the University of Aberdeen and made his landmark discoveries in tropical medicine when he was a surgeon in the Chinese Customs in Amoy.[65] Both services also actively recruited graduates from the University of London, whose career prospects in England were less secure than those of the Oxbridge elite. Newly recruited members, approximately twenty years of age, spent their first months in China cramming in the language before being sent to the various posts. Each service also employed scores of Chinese clerks on the staff as writers.[66]

These nonscientific organizations outstripped any scientific bodies in providing talent for research into the natural history of China. Only French missionaries, who had their own network and enterprise, and

Russians, who had unique access to the northern and northwestern parts of the Chinese empire, contributed as much.[67] Energetic and educated, junior members of the British Consular Service and of the Chinese Maritime Customs could make ideal naturalists.[68] Their positions also provided them with means and opportunities for gathering scientific data. Some of them had already acquired an interest in natural history at home: Robert Swinhoe, Chaloner Alabaster, G. M. H. Playfair, and Edward C. Bowra belonged to this category.[69] The majority took up the hobby in China. They were excited by the vision of infinite opportunities in a vast empire whose flora and fauna were scarcely known to Westerners. The more enthusiastic recruited their friends and colleagues. While stationed in Fujian, southern China, Swinhoe guided his consular colleague H. F. W. Holt to ornithology and lent the beginner his Chinese taxidermist.[70] Augustine Henry, a Customs officer and zealous botanical collector, encouraged his colleague Hosea Morse and Morse's wife to pursue botanical collecting.[71]

Although few of these naturalist-collectors enjoyed a scientific reputation, they had much to gain in the pursuit. They collected for fun, for curiosity, for self-improvement, for ambition, for the intellectual satisfaction of making a contribution to science, for the good feeling of having a respectable hobby; and sometimes they were requested by friends or ordered by superiors to investigate certain products of nature. Henry began botanizing in Yichang (Ichang) partly out of boredom. The life of Westerners in minor treaty ports could be depressingly routine and unexciting. Collecting specimens joined tennis, card games, and hunting as regular pastimes.[72] Some of the young diplomats were typical Victorian mountaineers and loved a good hike. William Hancock had scaled high peaks in Europe and the Americas, collecting along the way botanical specimens for Joseph Hooker (who was himself no faintheart and had breathed the thin air on top of the Himalayas).[73] While in China, Hancock collected for Hance and Hooker and, true to the Victorian spirit of "muscular imperialism" and following the example of Swinhoe, he visited the alleged head-hunting tribes of Taiwanese aborigines.[74]

Overall, the administrations of the Chinese Customs and the British Consular Service supported the scientific activities of their members as long as the regular official work did not suffer. Yet, due to the different political natures of the two establishments, they held different official positions with regard to assisting British scientific organizations in research. Robert Hart instructed the commissioners of the Chinese Customs to investigate a series of subjects relating to fisheries, medicine, and economic botany, and he was respected by Western residents in China for his "enlightened" support of science.[75] But he never offered much official assistance to Kew Gardens, in spite of Kew's direct requests. Impressed with Augustine Henry's collections, Thiselton-Dyer suggested that Hart grant the young man a long leave for botanical fieldwork. To his disappointment, however, Henry received only a three-month extension of his regular summer vacation.[76] Hart did not explain why he soft-pedaled any cooperation with Kew Gardens; he might have found it problematic and unjustifiable to submit to a British organization the resources and facilities of a branch of the Chinese government.[77] Nevertheless, he did permit indirect assistance and private actions. Many years later, after taking stations in Hainan, Taiwan (Formosa), and Yunnan, Henry suspected that he had been posted to these remote and little-explored places because of his interest in botany.[78] And the ornithologist John David Digues La Touche also obtained from Hart leave to do fieldwork in the Wuyi Mountains (southeast China), for which a legless lizard discovered there was dedicated to him.[79]

The matter was very different with the British Consular Service in China. The establishment belonged to the British government and felt the force of Hooker's political clout. Kew Gardens submitted their requests directly to the Foreign Office in London, which then passed them on to the Consular Service in China. The Service's headquarters in Beijing usually complied and commanded H. M. Consuls in China to execute the orders. Most of the investigations concerned economic botany. Because some of the consular officers were interested in natural history and had privately been corresponding with Kew Gardens, they often went the extra mile in inquiring into the matters for Kew.[80]

Scientific Networks

Naturalists whose reason for being in China was not primarily scientific—civil servants, missionaries, and merchants—used research methods that depended on, and were determined by, their duties. Both the Consular Service and the Maritime Customs, for example, reshuffled their staffs frequently, and their officers were transferred from station to station every few years, sometimes more than once a year. This internal institutional policy strengthened the naturalists' networks and furthered their information collecting. It provided them with mobility, social contacts, and promising new fieldwork sites. During his career, a consular officer might be appointed (as was Robert Swinhoe) to ports as far apart as Danshui (Tamsui) in Taiwan and Yantai (Chefoo) in Shandong, and several others in between.[81] The Maritime Customs adopted very much the same program. Every move meant an opportunity for working at a new field site, especially if the post was in a place then little known to Western science.[82] Swinhoe amazed naturalists in Europe with the numerous zoological discoveries he made in Taiwan in the 1860s, hitherto a terra incognita to the European scientific community. When he was a vice-consul there, he lived in a modest Chinese-style house at the port of Danshui, northern Taiwan, whose entire foreign population amounted to four souls.[83] He had little official business, and he devoted much of his time to natural history. A couple of years later, he was removed to Takow, southern Taiwan, equally unknown to science, where he made many more discoveries.[84]

Frequent transfers also promoted the social contact and private networking among the officers. In the course of his career, an officer could establish friendships with many colleagues who also moved from place to place, and could thereby forge an extensive network of correspondence across much of China proper. Many officers shared an office at one place or another at some points during their careers. Friendships were kept up through regular correspondence. The treaty ports were connected together through sea routes, waterways, and serviceable roads. Since the 1860s, dozens of steamers regularly sailed on the Yangzi between Hankou (Hankow) and Shanghai, a distance of six hundred miles; and after 1878, the service extended to Yichang, a recently established treaty port, a further four hundred miles upstream. Courier services were speedy and reliable. A letter sent from Hance in Canton dated April 1, 1885, reached Augustine Henry in Yichang on the twentieth of the same month. Another, dated June 7, arrived in Yichang on June 27.[85]

Hence, although the British resided in Chinese towns and cities scattered hundreds and even thousands of miles apart, they formed a close-knit network with a few particularly concentrated communities like the one in Shanghai. Because of this institutional formulation, an enthusiastic naturalist could influence quite a few of his colleagues, encouraging them to take up natural history or persuading them to collect specimens and natural history data for him in different parts of China. The naturalists—be they homebodies like Henry Hance or intrepid explorers like Robert Swinhoe—had little difficulty in building networks of scientific information, which, as we shall see in Chapter 4, were further extended by the many Western journals and newspapers published in China.

Those who enjoyed a reputation among Westerners in China for their expertise and achievements as naturalists acquired new specimens from willing collaborators. Known to his colleagues as one who neglected his wife for zoology, Swinhoe likewise received animals and information from many unexpected sources.[86] Whenever and wherever a Briton in China got an unusual animal, he was likely to send it to Mr. Swinhoe. A consular officer in Shanghai brought back from Japan a pair of "cranes" for the garden of the Consulate, but the Japanese newcomers turned out to be aggressive and attacked the Chinese cranes already there. The bullies were immediately banished from the garden and dispatched to Swinhoe, who found them to be not cranes but a new species of stork.[87] The senders could be rewarded by naming the animals or plants after them. Swinhoe asked Henry Kopsch of the Chinese Customs to acquire specimens of a deer seen in his district and promised to name the deer after him if it turned out to be a new species.[88] For a similar reason, William Gregory, a vice-consul, took great trouble to send Swinhoe "an enormous Pipe Fish" from Taiwan.[89]

However, the naturalists' duties as government officials also limited their natural historical research. Time constraints seemed to be a common problem. The officials were supposed to keep office hours from nine to four except Sundays, and their workload could be heavy during certain times of the year. Complaints about their duties as government officers surfaced from time to time in the naturalists' letters to scientific friends in Europe. After explaining to Thiselton-Dyer the many official dispatches he had to write, a consul speculated that H. M. officials in China had more routine office work than did their colleagues in Britain.[90] Confined to the office, they could not perform fieldwork and pursue natural history as much as they liked, so many of them employed native collectors. Of course, the workload varied with the official's position and location. Duties at some posts were lighter than at others. By the middle of the nineteenth century, the glorious days of Whampoa as a port were long gone, and Hance had plenty of time for botany. The successes of Swinhoe and Henry tell a similar story.

In a foreign country like China, the naturalists necessarily sought the help of the natives, recruiting local Chinese magistrates, merchants, and others for specimens and information either through official channels or personal connections. The naturalists' positions as government officials provided them with status, authority, and respectability, all of which facilitated their scientific inquiries. Chinese officials did not necessarily like Western government agents, but they were usually polite toward them and were willing to assist their inquiries as a gesture of friendship. This kind of everyday social intercourse impinged only indirectly on Western imperial domination; it mainly followed the tradition of social etiquette between Chinese and British officials. Curious about whether the Manchurian tiger was a new species, Swinhoe secured a skull through the help of a Chinese governor.[91] On another occasion, he procured the skin of a rare bear "through the civility of a high mandarin."[92] Conversely, the animals that Swinhoe acquired by various means for zoological research invited requests from Chinese officials for the parts that were valuable in Chinese medicine. When he was in Taiwan, for example, he once acquired a mountain goat, and the "high mandarin of the town begged the blood of this animal of me, and esteemed the gift a great favour."[93]

For certain types of investigations, the Customs officers occupied a unique position because their duties required them to inspect the trade goods that trafficked through the ports. They had the opportunities to collect information about particular vegetable products and medicinals, identify the plants, and determine their origins. In addition, they had ready access to the Chinese who traded or refined these natural goods and who possessed firsthand information about the drugs, spices, fabrics, vegetable oils, indigo plants, and other botanical items of value. As a Customs officer in Canton, E. C. Bowra, "whose position gives him great advantages in inquiries of this nature," procured for Hance a living specimen of the China root, a drug whose source Hance "had long been anxious to ascertain."[94]

Bretschneider, an authority on Chinese *materia medica*, urged the Customs officers to make inquiries of this nature and wrote an essay describing dozens of important vegetable products that needed further investigation and evaluation of their economic potential.[95] Admittedly, only a few of the Customs officers knew enough about natural history to carry out thorough scientific research, and most of them had to be satisfied with gathering specimens and compiling information. Bretschneider criticized the many factual errors—particularly misidentifications of plants, drugs, and so on—contained in the research reports on silk and *materia medica* issued by the Chinese Customs.[96] We should be careful to interpret his complaints over missed opportunities, however, because he himself frequently consulted the very studies he criticized.

Research into *materia medica* might look quaint and marginal in our age of chemical drugs, but it directly contributed to the development of botany and pharmacology in the nineteenth century. In the 1860s, the London-based pharmacologist Daniel Hanbury undertook the daunting project of identifying various medicinal substances and analyzing their chemical contents. As many drugs had been imported or transmitted from one place to another, his research depended on an extensive network of correspondents across the world. His major correspondents in China included his brother Thomas, a merchant in Shanghai, and Hance. He once requested Hance to identify the plant that yielded the "lesser galangal," a drug or spicy stimulant imported from China. The Europeans had known the drug, in the form of dried rhizomes, for centuries, but its origin remained a mystery. In response to Hanbury's request, Hance mobilized his scientific network in China, and eventually several individuals contributed to the quest.

During an expedition on an official duty to the island of Hainan, Sampson came across a plant that Hance suspected to be the prize, but he missed the opportunity of collecting complete specimens. Luckily, a similar expedition took place in the following year, and its members included Edward Taintor, an American in the Chinese Customs Service, who had previously collected for Hance specimens of an important oak tree in north China. This time Taintor successfully brought back from Hainan five living specimens of the plant in question. For comparison, Hance asked one of his regular correspondents, Johannes E. Teijismann, Director of the Buitenzorg Botanical Gardens in Java, to send over specimens of a related plant. After examining the living plants, the drugs found in China, the specimens from the Dutch East

Indies, and the original sample from Hanbury, Hance concluded that the plant was new to botany. Hance and Hanbury then each wrote an article describing the discovery in the fields of botany and pharmacy, respectively. This example demonstrates how individuals in the European scientific community, institutions of the informal empire, and a European colonial botanical garden worked together to carry out a scientific inquiry in the overlapping fields of natural history, medical botany, and pharmaceuticals.[97]

Missionary and Commercial Institutions

Although the majority of British naturalists in China were consular or Customs officers, there were also a number of missionaries notable for their exertions in natural history. Compared with Catholic (especially French) missionaries, Protestant missionaries in China, mostly British and American, contributed little to natural history. The achievements of several French missionaries in China during this period attracted the full attention of the scientific communities in Europe—for example, celebrated zoologist Armand David, conchologist Pierre Heude, and botanic collector Jean Marie Delavay.[98] No Protestant missionary accomplished half as much as they did. The marked difference between the Protestant and the Catholic missionaries' achievements in natural history stemmed from their different approaches to missionary work. Catholic missionaries had long established themselves in the interior and maintained an extensive network across the Chinese empire. Spread out and stationed in their designated places for decades, the members dressed and lived as the natives did. They enjoyed unique access to western and southwestern China, including Yunnan and the Sichuan-Tibet border, where the flora and fauna were abundant and peculiar. The French, who retained some elements of the early Jesuit approach in China, which emphasized Western achievements in science, established museums in Beijing and in Xujiahui, near Shanghai. Much of David's and Heude's scientific research was conducted in association with these museums.[99]

The Protestant missionaries, on the other hand, were usually married and accompanied to China by their wives and families. They mainly stayed in coastal cities, preserved their Western lifestyle, and devoted their attention to serving the urban and suburban poor.[100] The work of the China Inland Mission, a movement that aimed to extend Protestant missionary work into the interior, gathered momentum only in the closing decades of the nineteenth century.[101] Confined to the cities, the Protestant missionaries did not have easy access to good fieldwork sites. Besides, the treaty ports had been the places most heavily researched by Western naturalists, so it was difficult to find novel flora and fauna in their vicinities. "It is not surprising that English missionaries have done so little for the collection of Natural History from China . . . ," a Catholic missionary explained, "they are mostly confined to towns, having residences there."[102]

It was a fair comment, but one cannot therefore infer that Protestant missionaries in China had no interest in or simply ignored natural history. A number of them—D. J. McGowan, B. C. Henry, S. Wells Williams, Alexander Williamson, John Ross, John Chalmers, Walter Medhurst, Sr., and Ernst Faber, for example—collected for scientific institutions, such as Kew Gardens, and for their scientific friends in China.[103] Some, such as John Ross, traveled for missionary work and collected along the way, and others, notably Frederick Porter Smith, studied Chinese *materia medica*, particularly because the Protestant missionaries were very active in medical missionary work and had founded several hospitals in China.[104]

The largest group of British residents in China were neither diplomats nor missionaries, but merchants (including engineers and other professionals working in commercial enterprises).[105] Despite its size, this group was less visible in natural history than were the diplomats and the missionaries. This difference was, above all, a consequence of the overall low interest in natural history among the merchant class. But there were also external factors. The vast majority of the merchants resided in major treaty ports, for obvious reasons. Big business could be found only in big cities. Like many missionaries, they flocked to geographic locations unfavorable to fieldwork of any kind. Nevertheless, a few did turn their attention to natural history. John C. Bowring, son of Sir John, the savant governor of Hong Kong, pursued a mercantile career and amassed a large collection of insects by the end of the 1850s.[106] Thomas Kingsmill, engineer and merchant, was one of the few British geologists in China and made good collections of fossils, many of which had been found at drugstores in Shanghai.[107] Francis Forbes, head of a large American company, took up botany and collecting.[108]

The merchants made their greatest impact upon zoology, in part because hunting was a highly popular pursuit among Western residents in China. Together with La Touche, a Customs commissioner, the merchants Frederic William Styan and Charles Boughey Rickett more than any other Britons filled the gap left by Swinhoe's early retirement in 1875. The merchants also sponsored geographical and geological investigations into China's natural resources, trade routes, and market potential. The most notable attempt was Ferdinand von Richthofen's survey of China's mineral riches in the 1870s. Eager to assess the natural resources of different parts of China, Western (predominantly British) merchants in Shanghai employed Richthofen, a German geologist, to examine the geological formations and deposits of coal and other minerals. Richthofen traveled all over China proper, observing its geology and geography.[109]

Trade sustained the flourishing British communities in China, and the major trade items were products of animal or plant origin, notably tea, oils, and plant fibers. The successive curators of the Shanghai Museum, a natural history museum founded in 1874 and affiliated with the North China Branch of the Royal Asiatic Society, all noted the importance of enlarging its collection of products of nature and artifacts in trade.[110] Later in the century, a plan for a "Trade and Commerce Museum" was presented.[111] The museum would display artifacts and natural products in trade and would style itself as a window on the China trade. The museum would therefore advance commercial, scientific, and educational goals. Ultimately, the desire of British merchants in China to protect their existing commercial interests might have prevented them from doing more in investigating the "secrets" of many Chinese products of vegetable or animal origin. As traders, they naturally wanted to have a monopoly on supplying Western markets with these products, and letting the Chinese keep the supply actually worked better for them than letting it shift to the colonies. The development of tea production in India would eventually damage the profits of British tea traders in China.

Consuls, Customs officers, and missionaries, on the other hand, had relatively little personal investment at stake and freely inquired into economic botany and related matters. Even a high-minded taxonomist like Henry Hance plunged himself into intensive research on wild silkworms, matting grasses, and other subjects in economic botany.[112] Rutherford Alcock of the Consular Service not only procured David's

Science and Informal Empire 83

deer, a recent discovery, for the Zoological Society of London, but also sent them Chinese sheep, noted for their amazing reproductive power, in the hope of introducing them into Britain.[113]

Although the diplomatic, the missionary, the commercial, and the botanical establishments discussed here differed in their manifested purposes, they shared important similarities. They were all involved in collecting, processing, and distributing information about China. The Consular Service and the Maritime Customs were, respectively, British and Chinese government agencies dealing with diplomatic relations (defined broadly) and trade. They constantly kept their eyes on the political developments, social change, and economic opportunities in China. The missionary societies studied the Chinese with the aim of converting them. The merchants eagerly collected information about trade routes, potential new markets, and profitable export products. The Hong Kong Botanic Gardens, though hampered by bureaucratic obstacles, struggled to achieve the research goals that Joseph Hooker laid out for it—investigating the flora of China and the origins of Chinese vegetable products. As the major functions of these establishments included collecting, analyzing, circulating, and disseminating information—about China, about its people, culture, society, commerce, geography, and natural history—they developed social networks and an infrastructure to carry out the tasks. Reliable channels of communication carried private correspondence as well as official dispatches fast and wide. Thus equipped, these institutions operated independently or together to gather relevant information and produce factual knowledge about China. Inquiries into China's natural history constituted part of this enterprise of "surveying" China.

Scientific Imperialism and Factual Knowledge

Research into China's natural history usually involved the natives. In pursuing fieldwork, for example, the naturalists typically hired native guides and collectors, and invariably sought assistance from the local people, who possessed unique information about the plants and animals of the place. For inquiries into economic botany, reliance on the native Chinese was almost inevitable. They controlled the supply of trade items, and they knew what plants and animals the products came from and where those plants and animals could be found. The investigations required more than simply identifying the origins of the products. It was also necessary to find out the methods used to process and refine them. Because the Chinese occupied certain links in the networks of information, one of the main tasks of the naturalists was to secure their cooperation.

Kew Gardens was keen to find out the origin of an unusual kind of "camphor oil" traded by way of Hainan. There had been confusion as to its origin, for it was extremely difficult to identify the plant from the final product. Consul M. F. A. Fraser in Hainan took advantage of his position and asked around among the Chinese for information. Some Chinese traders from Guangxi, the province where the oil was produced, were familiar not only with the plant in question, but also with the method of distilling the oil. They pointed out to Fraser the correct plant, and upon request, they explained the process of manufacturing the oil.[114]

One of the most significant cases of acquiring indigenous knowledge of manufacturing plant products was probably Robert Fortune's mission to recruit Chinese tea manufacturers for the East India Company's tea plantations in India. While in the tea districts of China, Fortune was granted permission to observe the process of manufacturing tea on many occasions, and he wrote down the details as carefully as he could. Yet many of the skills and techniques could be mastered only by years of experience and practice; this kind of tacit knowledge easily eluded even the most attentive observer. With the help of the Consular Service, Fortune succeeded in persuading quite a few Chinese tea manufacturers to emigrate, bringing with them the implements of their trade, to help the British develop tea plantations in India. It proved to be a shortcut to successful technology transfer. The tea trees transplanted from China perished in India, and it was the native wild tea trees and hybrid types that finally made up the output of India teas, but the knowledge of manufacturing tea had come from Chinese immigrants in India and the tea manufacturers who went to the tea plantations.[115]

It would be imprudent to share one's trade secrets with one's business competitors, and the Chinese were not oblivious to that simple fact. The naturalists' impressions of their informants differed, but many cited the long-standing Western belief in the Chinese suspicion and jealousy of foreigners.[116] Whether it was their own imagination or not, the naturalists often felt that the Chinese withheld vital data. When the seeds they acquired from the Chinese did not germinate, they suspected the Chinese of having boiled, baked, or poisoned them to protect their own trade interests. This notion of Chinese operating in bad faith was widely held by the British. In a letter to the *Gardener's Chronicle*, John Henslow thundered, "I have more than once received seeds from China in these little jars, but scarcely any of them germinated, and I believe it is a custom with the Chinese to scald the seeds they sell to barbarians!"[117]

Determined to solve the puzzle of the Chinese seeds, Fortune called upon Aching, the nurseryman mentioned in Chapter 1, for a demonstration of how seeds were selected and packed for export. Poor old Aching had "got a bad name" among the British over the years because the seeds he sold them often did not germinate.[118] After examining the procedure, however, Fortune decided to give the Chinese some credit for honesty: "The Chinese are certainly bad enough, but, like other rogues, they are sometimes painted worse than they really are."[119] In the end, he attributed the problem to long-distance transport. Nevertheless, doubts continued. After several failures in growing the star anise, a plant of commercial value, in the 1870s, Hance was convinced that the Chinese baked the seeds before parting with them. But his friend Ford found out later that the same kind of seed sent to him by the French in Tonkin did not germinate, either. Ford then concluded that the seeds easily lost their vitality.[120]

If these examples suggest the naturalists' inclination to distrust the Chinese, the suspicion was probably mutual. It should surprise no one that some Chinese might have assertively guarded their profits. In 1882 Ford traveled a long way to an area where valuable cassia trees grew, only to learn that the Chinese there would not let him have more than a few living plants. They were reluctant to share their money-making trees with foreigners. Ford later must have congratulated himself on his tactics. He sent one of his Chinese collectors, disguised as a dealer, into the cassia district farther inland to "buy up all the young plants in the nurseries," before the natives discovered his true identity.[121] During his journeys in China, Robert Fortune came across a "mosquito tobacco" that proved to be wonderfully effective in warding off the pests, so he intended to find out the ingredients of the drug. He called upon a shop making and selling the item, asking prying questions about the ingredients, their relative proportions, the process of manufacturing, and so on, but received only evasive answers. Fortune was not ready to give up, however; he put the matter in the hands of his Chinese associates, who finally obtained the valuable information after months of insinuation.[122]

Whether these and similar accounts are reliable or not, they reveal a shared common strategy of representation used by the naturalists in relating their encounters with the natives. If the Chinese tried to guard their profits or used questionable business tactics, they were mean, jealous, and cunning. If the British themselves were doing that, even breaking Chinese laws, they were portrayed as smart, adventurous heroes outwitting the conceited natives. Robert Fortune's highly popular travelogues were full of episodes of this kind. This kind of orientalist representation allowed the naturalists to justify their otherwise less-than-honorable actions.[123] The narrative was slightly complicated by the presence of the Chinese "helpers," but they were conveniently reduced to passive figures in compliance. In this type of narrative, the Chinese were either stripped of their motives and initiative or, if they displayed such traits, were cast as rogues or troublemakers.

Following recent scholarship, one might also be tempted to interpret the actions of the Chinese within the postcolonial model as resistance to imperial domination.[124] The Chinese were using "weapons of the weak" (for example, "sabotaging" the seeds, withholding information) to resist the dominating Western imperial power.[125] This reading is based on a model of power relations developed to interpret experiences in the colonies, and, valuable as it can be, it works less well in the context of China, which retained much of its political autonomy despite pressures from the West. In fact, the model may be too rigid even for most colonial situations.

It is of course necessary to consider the order of power relations that Western imperialism ruthlessly tried to impose on China by means of military force, economic aggression, and diplomatic pressure. Yet it is equally important to ask how the Chinese might have regarded the encounters in economic botany; for they, too, were agents of historical process and acted according to their interpretations of events. What did the Westerners' insistent inquiries *mean* to them?[126] Without direct evidence, one can only conjecture. In all likelihood, however, some

Chinese did refuse to share their trade secrets with Westerners (and, no doubt, with many other Chinese as well) for fear of aiding business competitors who would hurt their trade. They probably did not see the situation in terms of imperial domination and native resistance, but, rather, were following their business savvy.

Although the naturalists typically cited suspicion and jealousy to account for the difficulty in procuring reliable information from the Chinese, they accused the Chinese of another, even more fundamental, "sin" against the pursuit of truth. They claimed that the Chinese knew not the value of *facts*. Few of the faults of the Chinese ranked as high in the Western eyes as their (alleged) carelessness about "facts." It was considered to be one of the conclusive proofs of Chinese sloppiness, ignorance, and backwardness. It particularly annoyed the naturalists, who took pride in their relentless pursuit of "objective" data. Robert Fortune criticized the Chinese on this point only three pages into his first travelogue on China: "I have been often much annoyed with this propensity of theirs [carelessness about correct information] during my travels in the country."[127] Fortune was by no means the most unsympathetic of all Western commentators on the Chinese. But he repeated the same point again and again throughout all his books on his China travels: "[one] is very apt to be misled by the Chinese; not, perhaps, so much intentionally as from ignorance or carelessness as to whether the information given be correct or otherwise."[128] One can almost imagine Mr. Gradgrind from Dickens's *Hard Times* shouting at a Chinese, "Now, what I want is, Facts.... Stick to Facts, sir!"

Scholars have studied the formation of the modern conception of "facts" and described how this belief in factual knowledge gradually attained an indisputable cultural authority in early modern Europe.[129] Westerners in China had no doubt about the superiority of this knowledge tradition over whatever tradition the muddle-headed Chinese had. The best example of this attitude may be found in the management of the Chinese Customs, which was a Western-style institution, headed and run by Westerners, but belonging to the Chinese government. The British insisted on having full control over the management of the establishment, partly on the belief (and pretense) that the Chinese could not do it well. Robert Hart instructed his office to publish a series of impressively detailed statistical studies on important items going through the ports, including opium and Chinese drugs. In the same manner, they also put out volumes of scientific studies on silk, fisheries, public health, and so on. These volumes were compilations of reports sent by the officers from every treaty port to the headquarters. The officers evidently had made diligent efforts to collect, classify, and analyze the data. Claims of factual accuracy were manifest in the form and content of these publications: multiple-column tables and statistics, transplanted from Western practice, filled many of the volumes.[130]

It should be noted that the Chinese had their own tradition of gathering political, economic, and geographic data, and they, too, published them in official format. The *fangzhi* or gazetteer, an official publication issued by local governments, typically included "facts" of local history, geography, and natural history, and numerical data of population and revenue. Not only did Westerners in China know of them, they actually used them, mining them for information. However, they found fundamental flaws in the Chinese attitude toward facts. "My experience of the Chinese Empire teaches me," remarked Thomas T. Meadows, H. M. Consul and a prominent sinologist, "that it is impossible to obtain from Chinese sources materials for [statistical research]."[131] They were convinced that while the Westerners pursued objective, empirical knowledge, the Chinese inevitably conflated facts with fantasy, myth, and all sorts of misinformation. The Chinese loved grotesques; they were ridiculously unobservant and credulous; they could not tell facts from tall tales and did not care. Thus this orientalist discourse defined a Chinese otherness—their inability to appreciate the unique value of facts. In their general works on China, both John Davis and S. Wells Williams agreed that "the general state of sciences among [the Chinese]" reached at best the level in Europe "previous to the adoption of the inductive mode of investigation."[132]

As a result, Western naturalists believed that their mission to investigate the natural history of China and collect information about economic botany went beyond enriching the British Empire. Exploring the natural resources of China was justified on the grounds that it was mutually beneficial. Their rationalization ran like this. The ignorant and unscientific Chinese suffered from their inability to take full advantage of their natural resources. Worse, their instinctual jealousy blinded them from seeing how they themselves could benefit from letting the "enlightened" Westerners extract and use the natural riches to everyone's advantage. The Chinese were foolish and selfish not to see the light and embrace this high-minded goal. And although they sometimes had means of manufacturing ingenious products, they lacked the knowledge—above all, scientific knowledge—to improve them. Hence the naturalists were convinced that they had the *right* and *responsibility* to obtain the knowledge of not only China's natural history in general, but specifically its animal and plant products of economic value. In the hands of the Chinese, they believed, such treasures would be wasted. This logic of paternal imperialism underlined much of the naturalists' conviction that they ought to have access to the information they needed, though, of course, there was also plenty of utter arrogance and sheer greed.

Knowledge and Empire

In the imperial context, the activities of natural history—mapping, collecting, ordering, classifying, naming, and so on—represented more than matter-of-fact scientific research. It also reflected an aggressive expansion of cognitive territory defined in particular cultural terms. The "discovery" of a new bird or plant—classifying it, placing it in the Linnean or other universal taxonomy, describing it in strict scientific Latin, representing it in Western pictorial conventions and techniques, turning living samples of it into material embodiments of abstract scientific concepts and specialized terminology, configuring its global distribution in rigorously defined diagrams—privileged a specific way of defining nature, facts, and knowledge. In the nineteenth century, scientific expeditions, whose core concepts and activities involved collecting, measuring, mapping, and traveling, and whose ultimate goal was to write the natural history of the globe with exhaustive comprehensiveness and precision, originated in part from a view of geography and nature coupled with European expansion and from an assumption of the right of "objective" European scientists to travel and observe other continents of the world. This conviction of a right to know, a right that was ideally not restrained by human boundaries—particularly, the boundaries drawn by the natives against European scientific researchers—derived its authority partly from a belief in the universal validity of factual knowledge.

Thus one of the major components of scientific imperialism was the ideology and practice of collecting information and producing knowledge—knowledge that claimed to be factual, objective, scientific, and definitive—about other parts of the world. It asserted an epistemological authority, and it pointed toward an ideal and belief in bringing the natural world, regardless of national and other human boundaries, under the lofty vantage point of truth. British naturalists in China participated in erecting this empire of information, empire of knowledge. Another major component of this empire of information was that the knowledge it produced was not simply "objective," but also *useful*. For the naturalists, taking possession of the data of China's natural history and economic botany was in the interest of the natives as well as of the Europeans. They deemed that they could produce useful, scientific knowledge from the information and that the knowledge—whether it was geology, economic botany, or some other science—would eventually bring material benefits to the Chinese. Possessing, in this view, would prove to be a generous way of giving.[133] The natives' will had no place in this grand vision of scientific commonweal.[134]

Given the vital importance placed on the amassing of information, it was no accident that Joseph Hooker insisted that "exploration" and "correspondence" be a crucial part of Charles Ford's official duties as Superintendent of the Hong Kong Botanic Gardens. Both activities were to collect data—information and specimens. Similarly, it was only natural that Hooker instructed Ford to transform the Gardens into "the channel of communication" of botanical matters on China and that he urged Ford to "enlist the co-operation" of people who resided and traveled in the China region. This vision was only partially realized because of the tension between British imperial institutions and, more importantly, the external political situations. The network of colonial botanical gardens could not easily multiply in a noncolonial region. Yet British scientific imperialism did not stop at the colonial boundaries, but expanded along the legal, political, and economic apparatuses of the informal empire, in whose shadow Victorian naturalists in China developed information technologies and carried out their research.

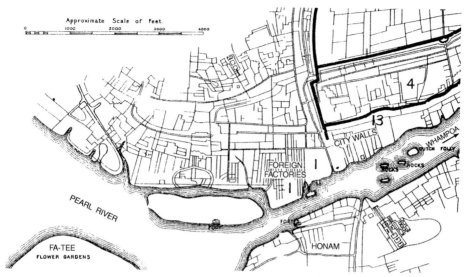

Figure 1. Vicinities of the Canton Factories. The Canton Factories were located outside the southwestern corner of the walled city of Canton, on the banks of the Pearl River. Although foreigners were not allowed to enter the city except on rare occasions, they were in daily contact with the Chinese along the Pearl River and in the neighborhood of the Canton Factories.

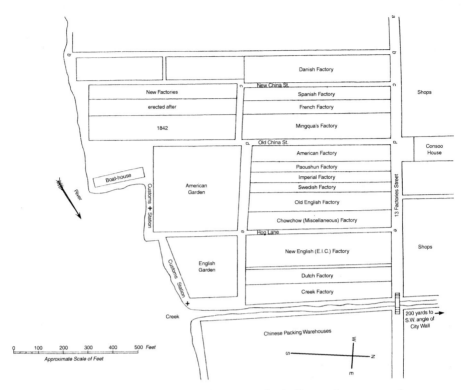

Figure 2. The Canton Factories. The Factories were basically warehouses rented to Western companies trading in Canton. The streets in the district were lined with shops and crowded with porters, peddlers, and foreign visitors.

Figure 3. A view of the Canton waterfront, c. 1770 (courtesy of the Victoria and Albert Museum).

Figure 4. Flower peddler. Plants and animals for sale on the streets of Canton supplied the British naturalists with many new discoveries (courtesy of the Victoria and Albert Museum).

Figure 5. Hong merchant's garden. British naturalists found horticultural treasures in the gardens of the Hong merchants (courtesy of the Victoria and Albert Museum).

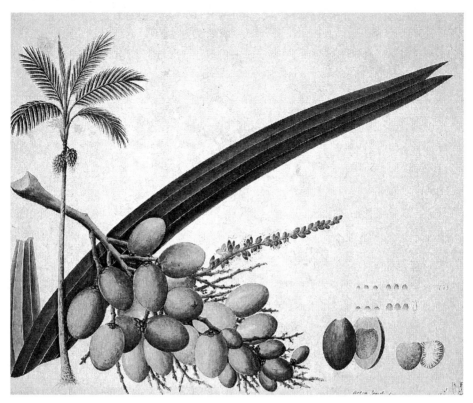

Figure 6. The betel palm. Chinese artists of export painting successfully adapted their skills to drawing natural history illustrations like this one (courtesy of the Museum of Natural History, London).

Figure 7. The red plum tree (courtesy of the Museum of Natural History, London). Both the plum tree and the crane (Figure 8) were popular subjects of traditional Chinese bird-and-flower painting. These two drawings from the Reeves collection display typical aesthetic devices of that genre.

Figure 8. The red-crowned crane (courtesy of the Museum of Natural History, London).

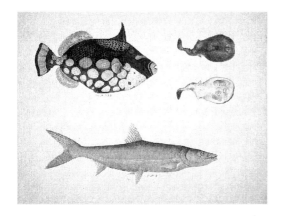

Figures 9a and 9b. Samples of Reeves's fish drawings (courtesy of the Museum of Natural History, London). In 1845, the ichthyologist John Richardson named more than 80 new species of fish solely on the basis of Reeves's collection of fish drawings.

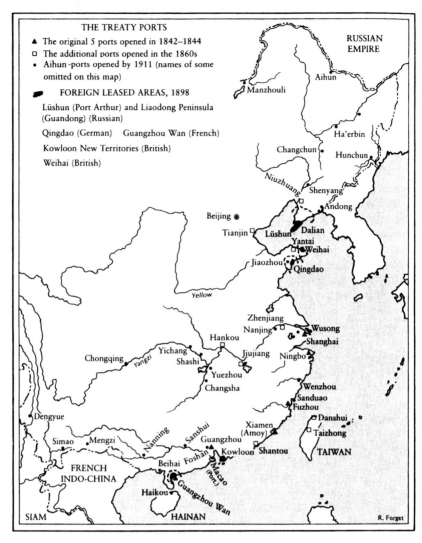

Figure 10. Foreign encroachments. Adapted by permission of the publishers from *China: A New History*, by John King Fairbank and Merle Goldman, p. 202, Cambridge, Mass.: The Belknap Press of Harvard University Press, Copyright ©1992, 1998 by the President and Fellows of Harvard College.

Figure 11. Henry Fletcher Hance, H.M. Vice-Consul at Whampoa and an authority on Chinese plants (courtesy of the Royal Botanic Gardens, Kew).

Figure 12. Robert Swinhoe, a member of H.M. Consular Service and an indefatigable field zoologist (courtesy of the Royal Botanic Gardens, Kew).

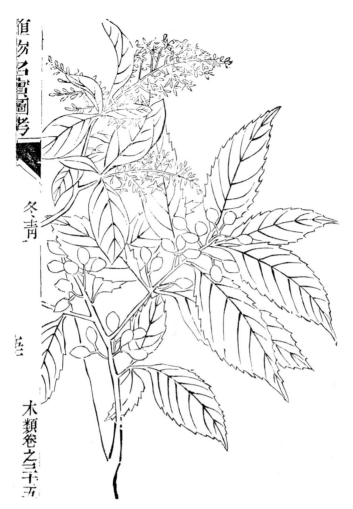

Figure 13. An illustration from *Zhiwu mingshi tukao* (1848), a pictorial dictionary of plants frequently consulted by Western botanists in China in the late nineteenth century (courtesy of the Needham Research Institute).

Figure 14. Tuo long (the *tuo* dragon). A. A. Fauvel used this illustration from *Bencao gangmu* (1596) for his research on the Chinese alligator. Ernst Mayr Library, Harvard University.

Figure 15. Frontispiece of a manual for Western sportsmen in China. Pheasant shooting was extremely popular among Western residents in Shanghai and the Yangzi Valley. Ernst Mayr Library, Harvard University.

Figure 16. Chinese hunters in the field. The matchlock was the standard weapon of a Chinese hunter. Ernst Mayr Library, Harvard University.

4

Sinology and Natural History

IT HAS BEEN argued that the second half of the nineteenth century witnessed an increasing professionalization of science in Britain. The growing dissatisfaction with the traditional establishment of science, which seemed to be scientifically stagnant and politically conservative, brought about calls for institutional reform, professional autonomy, and intellectual revolution. This process of professionalization ushered in new approaches and areas of research in science. Although Joseph Hooker's primarily research interests were set on collecting, systematic botany, and botanical geography, his successor at Kew Gardens, William Thiselton-Dyer, promoted since the 1870s botanical sciences that depended on laboratory research, ecological concepts, and agricultural application.[1]

This narrative of professionalization can be misleading if we stretch it to include British scientific research in nineteenth-century China. New botany never became the main staple of that research. The difference between the British presence in China and in the British plantation colonies helps to explain why little was done to advance new botany in the China region. The British lacked the kind of intellectual and institutional support needed to expand certain areas of research in China. The ambition of imperial botany had to come to terms with the political reality and institutional constraints in a territory not directly ruled by the British government. During the second half of the nineteenth century, however, British natural history in China did develop new research provinces that drew strength from the increasing professionalization of a scholarly discipline, namely, sinology.

The institutionalization of sinology took place in continental Europe over the first half of the nineteenth century, and somewhat later in Britain.[2] Sinology was a latecomer to what has been called the discursive formation of orientalism.[3] In China, Western sinologists had scholarly organizations, the most important of which was a branch of the powerhouse of British orientalist scholarship—the Royal Asiatic Society. Like natural history, sinology must also be understood in context. Many nineteenth-century Westerners in China avidly pursued both sinology and natural history; they did not pack them in neatly separate intellectual compartments, however, but considered them as related enterprises. As we shall see later, the sinologist-naturalists shared with other orientalists certain interests, ideas, and research techniques, and their orientalist background informed their research into China's natural history.

The main goal of this chapter, however, is not to emphasize the ideology of European imperialism in nineteenth-century scientific representation of China's natural world.[4] Instead, it examines the ways in which Western research into Chinese writings about the natural world resulted in the formation of an (so to speak) interdisciplinary discourse between sinology and natural history. Our primary concern here is to describe and explain the historical circumstances and ways in which the Western naturalists approached, interpreted, and translated Chinese knowledge of the living world as contained in Chinese texts. The focus is on the process instead of the result. To understand the intellectual subtleties of knowledge translation in a particular historical condition, we must examine, in this case, why the naturalists consulted Chinese works, what Chinese works they most frequently used and why, and how they construed and unfolded Chinese texts for the purpose of natural historical research.

Because of its sinological content, the work of the sinologist-naturalists may appear distinctive, a tiny island of eccentric scholarship in the grand river of Western natural history—isolated, localized, and remote from the mainstream. I will argue, however, that at a deeper level their work was a logical outgrowth of the tradition of natural history; the use of textual evidence, philological data, and historical documents had always been part of the scholarship and practice of natural history. The tradition of textual practice—as opposed to other forms of scientific practices, say, fieldwork and museum practice—has been neglected by historians of nineteenth-century natural history.[5] This chapter challenges the conventional view that the role of texts diminished in natural history at some point in the early modern period and that the science metamorphosed into one that was based on direct observation and anatomical investigation.[6] I suggest that textual practice and the tradition of employing humanistic scholarly apparatuses remained fundamentally important in that branch of science as late as the nineteenth century.[7]

Much of our discussion concerns cross-cultural translation between different knowledge traditions and between different languages. Some scholars have asserted that there was a fundamental cultural incommensurability between China and the West that was rooted in mental categories, languages, or worldviews.[8] Part of their intention was to critique the pitfall of viewing Chinese culture from a Eurocentric standpoint. Ironically, however, they ended up repeating the orientalist discourse by redrawing a demarcation between China and "the West."[9] In this view, each culture becomes a fixed, rigid structure, and different cultures are seen as mutually exclusive.[10] This viewpoint assumes and essentializes culture rather than explains it. Moreover, it ignores what historical actors actually *did* and turns them into puppets imprisoned in the glass house of culture.[11] The rigid dichotomy inherent in such a view makes any nuanced explanation of cultural encounters impossible. However complex the translation and comparison between Chinese and European knowledge about nature's objects may appear to us as modern investigators, the historical actors practiced it anyway, and it is exactly this phenomenon that we must explain within its historical context. By focusing on the textual practice and the naturalists' strategies of reading, I will examine the historical actors' motivations, goals, criteria, and practices of translation. I am interested in what they thought and experienced and how they defined, negotiated, and crossed the boundaries between knowledge traditions.

Extant documents do not allow us to reconstruct fully the roles and intentions of the Chinese participating in the scientific enterprise. The Chinese, no doubt, took part in the collection, interpretation, and translation of knowledge in research into Chinese books for natural historical research, though we probably will never, unfortunately, be able to ascertain the details of their involvement in the enterprise or to examine how they negotiated linguistic and knowledge categories with their Western associates in translating Chinese works.[12] Without native assistants, even the best Western sinologists in China found themselves skating on thin ice.[13] The missionary James Legge, one of the greatest Victorian sinologists, depended heavily on the expertise and learning of his assistant Wang Tao in writing his celebrated *The Chinese Classics* (1861–1872).[14]

Another historical contingency also determined the terms of Sino-British contact in natural history. Throughout the nineteenth century, the British naturalists never encountered Chinese scholars who were particularly learned in natural history. They never met, for example, Zhao Xuemin, who wrote a supplement to the sixteenth-century herbal *Bencao gangmu* in 1765; Li Yuan, who produced an encyclopedia on animals in 1816; or Wu Qijun, whose monumental *Zhiwu mingshi tukao*, the "Pictorial Treatise on the Names and Natures of Plants," published in 1848, would become a standard reference work for Western botanists in China in the late nineteenth century. We can only imagine what results such an intellectual encounter might have brought about, yet we do know that the European naturalists extensively mined the Chinese literature on the natural world.

Early Attempts

The connections between sinology and natural history began long before the nineteenth century. During the early modern period, Jesuit missionaries, who had established themselves in China since the late sixteenth century, played a key role in transmitting to Europe information about China, including its literature on the natural world.[15] But few Europeans could read Chinese. Matteo Ricci claimed from China that "no other language is as difficult for a foreigner to learn as the Chinese."[16] With rare luck, seventeenth- and eighteenth-century scholars in Europe might find visiting Chinese to help them crack the script.[17] Almost all of them, however, had to work solely from the sundry ideographs recorded in Jesuit writings.[18] So exotic and esoteric did Chinese seem to Europeans that many theories were generated to explain its origins and nature. Some, like John Webb, believed that it was the first spoken language of humankind.[19] Others, including Athanasius

Kircher and later Joseph de Guigness and Lord Monboddo, were convinced that it derived from Egyptian hieroglyphs.[20] In the international search for a universal language during the seventeenth and eighteenth centuries, John Wilkins and Gottfried Leibniz, among others, looked for inspirations in Chinese.[21] Issues about Chinese popped up again and again in the Enlightenment debates about the origin of language. Yet, with little information to count on, European scholars ended up scratching their heads and performing intellectual acrobatics.[22]

The interest in China's natural history and in Chinese literature converged in many ways. The Chinese had produced an enormous amount of literature that dealt with plants, animals, and minerals, and the Jesuits made good use of it. As early as the mid-seventeenth century, Michael Boym, a Polish Jesuit missionary to China, wrote works on Chinese drugs and medicine that were based partly on Chinese herbals.[23] His writings were further incorporated into Kircher's influential *China illustrata*, which gave rise to debates over the musk deer and the "snakestones."[24] Ginseng, rhubarb, camphor, and other curious plants and animals purportedly of Chinese origin continued to baffle generations of European scholars, not least because of scanty and often contradictory reports.[25] Even Linnaeus was forced to grope in the dark. Knowing little about tea, he was eager to introduce and grow the plant in the soil of Sweden. His great scheme flopped badly.[26] The volumes of du Halde served generations of European naturalists as a storehouse of information about Chinese flora and fauna.[27] The work was translated into all major European languages. It contained excerpts from the great Chinese herbal *Bencao gangmu* and made it well known in the West.

In the nineteenth century, both sinology and Western research into China's natural history took on new forms. Sinology was becoming institutionalized: A chair in Chinese language studies was created at the Collège de France in 1814;[28] other European countries followed suit in the subsequent decades. Most of the first academic sinologists in Europe had never set foot on Chinese soil. As the century unfolded, more and more Westerners went to China, many residing there for years or even decades before returning to Europe. In the later part of the nineteenth century, academic sinologists in Europe were mostly recruited from this pool of talent and experience. Meanwhile, the widening of Sino-Western contact in the wake of the Opium Wars brought about new opportunities for Westerners to study the flora, fauna, and geology of China.

Before 1800, only a few Britons acquired any acquaintance with the Chinese language. The situation improved slightly during the first decades of the nineteenth century when several individuals, notably George T. Staunton and Robert Morrison, made headway into sinology and earned grudging recognition from their continental colleagues (see Chapter 1). Many of these pioneering British sinologists and their colleagues were also interested in natural history. They were at the same time tantalized by the reputed natural bounty beyond their reach and frustrated by the impossibility of gratifying their curiosity. "[The] works of nature in China are shut out from our gaze," sighed a Western resident in Canton.[29] Without direct access to the interior, the naturalists sought to maximize whatever they could lay their hands on; they hoped to open a window: "we can look into the books of the Chinese." "Their medical and botanical treatises are numerous and voluminous indeed, and we might reasonably promise ourselves a reward in reading them."[30]

However, it was only in the second half of the nineteenth century that many Westerners were able to look through the window. The number of Westerners who knew Chinese increased spectacularly between 1850 and 1900 as diplomatic and missionary organizations expanded. Few merchants bothered to learn Chinese; they depended heavily on the linguistic assistance of Chinese compradors and clerks.[31] Diplomats and missionaries, on the other hand, needed some proficiency in the language to carry out their duties. Both the British Consular Service in China and the Imperial Maritime Customs of China required their employees to learn to read and speak Chinese.[32] Indeed their employees' chances of promotion were determined to a great extent by their linguistic proficiency.[33] Like diplomats, missionaries in China needed knowledge of Chinese to advance their cause. Not only did they regularly have to use Chinese in their services, but they were also eager to gain insight into the heathen mind and to understand Chinese thought. Under these demands, it is little wonder that many missionaries became diligent sinologists.[34] These nonscientific organizations, as we saw in Chapter 3, also produced corps of researchers on natural history and provided networks for the gathering and dissemination of scientific information.

The Sinologist-Naturalists

After the 1850s Shanghai replaced Canton as the principal international port of China and had a thriving Western community.[35] As the population of Western residents in China grew, their cultural and intellectual life flourished. The Shanghai Literary and Scientific Society, founded in 1857, would become the North China Branch of the Royal Asiatic Society (NCB) in the following year.[36] It was the center of sinological research in China, holding regular meetings and possessing a large research library. Natural history ranked high among the intellectual pursuits promoted by NCB. In his presidential address, E. C. Bridgman, first President of the Society, urged members to pursue natural history and introduce the natural world of China to the West.[37]

The Society's journal was the most prestigious and influential Western journal in China. There were a few other learned journals that set themselves apart from the myriad Western periodicals in China. The major Protestant missionary journal *Chinese Repository* began in 1832. It published articles on a wide range of subjects—religious, historical, philological, ethnographic, and scientific. This important journal, however, ceased publishing in 1851. The *Chinese Recorder* filled the gap in 1867. *Notes and Queries on China & Japan* was also launched in 1867 and would evolve into *China Review*, a respected journal in Chinese studies. These were among the regular venues in which Western scholars in China could express opinions, discuss research, and announce discoveries. The *Journal* of the NCB, for example, published research articles of over one hundred pages as well as short reviews. *Notes and Queries*, as its title indicates, served mainly as a bulletin for serious queries and replies dealing with Chinese and Japanese subjects, but it also carried learned papers.[38]

The influence of these journals and organizations was felt far beyond Shanghai. The membership of the NCB, which reached 150 in the late 1860s and grew to some 250 in the 1880s, spread widely to other treaty ports and beyond. As for the journals, *Notes and Queries* hailed from Canton and Hong Kong; the *Chinese Recorder* was from Fuzhou, also in south China. The journals found readers even in Europe. The expansion of Western communities in China and the proliferation of channels for scholarly discussion in the 1860s and 1870s helped to form a discursive forum guided by the aim of sinological research.

This institutional development coevolved with the formation of an identity corresponding to the sinologists' expertise and geographical position. Sinologists in China saw themselves as experts with unique access to their research subject. Major contributors to the China journals enjoyed scholarly reputations not only among Westerners in China, but also in the international circles of sinology. Some of them, like the missionary James Legge, were among the foremost sinologists of the West.[39] Compared with their colleagues in Europe, they had the advantage of having direct contact with the Chinese. They could "investigate, at the fountain head, all that bears upon the physical, intellectual, and moral condition, of this countless population."[40] Moreover, it was infinitely easier for them to obtain help from native scholars. Not only did they have Chinese tutors teach them the language and Chinese writers assist them with translation, but they could even obtain native scholars' help through friendship or semi-official relationships. When W. F. Mayers, eminent sinologist and Vice-Consul at Canton in the 1860s, conducted research on the history of maize in China, he asked local Chinese mandarins for information and documents. And the Chinese Intendant of the Grain Revenue for the province of Guangdong prepared for Mayers a paper based on Chinese sources.[41]

Many of the sinologists took up natural history either as a pastime or as a serious pursuit. While few attained a reputation in Europe as distinguished natural historians, they were nevertheless lords of their chosen fields. In addition to necessary training in Western natural history, a competent investigator needed to possess a certain familiarity with Chinese literary traditions, knowledge of Chinese historical geography, and the ability to decipher the nomenclature and descriptions in Chinese works on natural history. As sinologists and as individuals who possessed firsthand knowledge of the enigmatic country, the naturalists took pride in their expertise. They wished to leave imprints not only in their writings, but also in their nomenclature.[42]

Swinhoe poked fun at "the great naturalists at home" for their indiscriminate application of the name *sinensis* to "everything that came to them from China." The specimens, he wrote, might have come "from some isolated spot in this vast empire, or from some distant part of Thibet or Manchuria."[43] Swinhoe himself could be very specific. He named a rat after the Chinese pirate-king Koxinga, who expelled the

Dutch from Taiwan and colonized the island, because the rat had emigrated from mainland China and dominated the indigenous species of rats in Taiwan.[44] He once proposed to name a bird discovered on the island of Hainan after the eleventh-century poet and essayist Su Dongpo, who spent many years on the remote island. The Cambridge ornithologist Alfred Newton did not appreciate this sinological touch and ultimately persuaded Swinhoe to give it up.[45] Less whimsical than Swinhoe, Emil Bretschneider, a Russian botanist in Beijing, recommended that plants be named after their indigenous names to preserve their origins. He cited the examples of Litchi, Moutan, and Yulan, with approval, and deplored the tendency to name plants after "savants or other persons (who frequently had nothing to do with the plant dedicated to them)."[46]

Intellectual Identity and Scientific Discourse

The establishment of an intellectual identity and a forum for exchanging information and expressing opinions also allowed the formation of a distinctive discourse, one that combined the vocabulary, concepts, and purposes of sinology and natural history. Although the China journals were not primarily scientific periodicals, Western naturalists in China found them a convenient place to publish their discoveries. The most accomplished, no doubt, preferred to have their technical papers appear in specialist journals in Europe. Both the botanist Henry Hance and the zoologist Robert Swinhoe, for example, each published scores of articles in leading scientific journals in Europe, including the *Journal of Botany* and the *Journal of the Linnean Society*, most of them reports in systematic botany and zoology.

Similar technical articles did occasionally appear in the China journals, yet, with some exceptions, notably Hance and the Jesuit zoologist Pierre Heude, most Western naturalists in China did not distinguish themselves in systematics.[47] A number of them, such as Swinhoe and the Lazarist Armand David, impressed their European colleagues as field naturalists.[48] Most were avid and able collectors who brought their sinological interest to bear on natural history. The scope of their interests ranged from economic botany, *materia medica*, or (to use an anachronistic term) ethnoscience—the study of Chinese knowledge of the natural world. They published in various forms of catalogs, travel journals, book reviews about the animals and plants of a particular area. Because their writings included sinological research—philological inquiries, discussion of Chinese history and culture, analysis of certain texts, and so on—most natural historians in Europe would find this part of their work esoteric, but could not ignore its often valuable conclusions. The sinologist-naturalists carved out a scholarly niche in which the discourses of sinology and natural history blended. Or put differently, they conversed together in a sino-scientific "creole" that would have defeated most naturalists in Europe.[49]

Overemphasizing boundary drawing, however, can obscure the fact that many sinologist-naturalists were actually well connected to the scientific communities in Europe. The enterprise of natural history still depended much on the network of correspondence so essential to the working of the nineteenth-century scientific community. Swinhoe, Hance, Bretschneider, Augustine Henry, and many other contributors to the China journals kept regular correspondence with scientific organizations in Europe. Naturalists in Europe had much to draw on their faraway colleagues' unique access to the natural history of China and their sinologically informed research. Western naturalists in China, for their part, gained prestige, influence, and indirect access to the well-stocked museums and herbaria in Europe.[50]

Sinological Inquiries and the Scope of Natural History

Like their colleagues in Europe, Western naturalists in China tended to specialize in one or two branches of natural history while dabbling in others. The sinologist-naturalists often preferred scientific research with a humanist bent, roaming in the border areas between natural history and civil history, between the natural sciences, such as zoology and botany, and the human sciences, such as philology and ethnography. Careful consideration of the sinologist-naturalists' reasons for consulting Chinese works may help us to fill in the truncated picture and reconfigure the map of knowledge in nineteenth-century natural history. The various reasons might be grouped into two rough categories: one primarily concerning the study of Chinese culture; the other, investigations of China's natural history.

As sinologists, the naturalists had noticed that the Chinese possessed an enormous amount of literature about plants, animals, minerals, and other natural specimens. Their herbals, agricultural and horticultural literature, encyclopedias, travel writing, and local histories all contained considerable information about the natural environment.[51] These writings formed a significant contribution to traditional Chinese literature.

The sinologist-naturalists recognized the value of this literature in understanding Chinese culture and sometimes approached the literature primarily from the standpoint of a sinologist. For example, Theophilus Sampson, a frequent contributor to *Notes and Queries* in the late 1860s, made clear in a paper that its subject was "figs . . . from a Chinese and popular point of view, and not figs in a botanical sense."[52] To assess Chinese achievements in natural history, Thomas Watters, a member of the British Consular Service, "devoted a considerable amount of time" to studying pigeons in Chinese literature; he also investigated the fox in Chinese myths.[53] In these cases, the naturalist-sinologists studied the native Chinese knowledge of their natural environment, and examined animals and plants in Chinese culture and society. In investigating Chinese writings about the natural world, sinologists hoped to understand Chinese civilization.

Exploring Chinese culture through natural historical literature constituted only a small part of the naturalists' enterprise. A more prevalent reason for digging into Chinese works on natural history came from their desire to unearth data that might assist investigations of the flora, fauna, and other natural treasures of China. The British faced considerable obstacles to any systematic exploration of China because of China's strict control of foreign activities in the interior. Travel restrictions would ease as the nineteenth century drew to its close, but random information brought back from forays into areas distant from the treaty ports could hardly add up to a satisfactory picture of the natural riches of so enormous an empire. Under these circumstances, British naturalists in China understandably wished to take full advantage of such a unique and potentially valuable source of information as Chinese literature.

Among the things that they hoped to discover in Chinese literature was information about Chinese *materia medica*.[54] Many Westerners in China turned their attention to Chinese works on natural history because of their curiosity about Chinese drugs and herbal remedies. They were convinced that there must be some efficacious items among the bizarre catalog of Chinese medicinal substances. Identifying the dried and processed roots, stems, and leaves was no easy task, for even an experienced taxonomist needed well-preserved flowers, fruits, and leaves to determine a plant's identity. In China, drugs came from all corners of the empire, especially from the mountainous regions in the interior, where few European naturalists had ever ventured and whose floras were little known. To trace an original plant from deformed, incomplete specimens necessarily involved much guesswork. The naturalists combed through the Chinese lore for clues.

In his pharmacological research, Daniel Hanbury (see Chapter 3) not only sought help from British doctors and naturalists in China, but also made efforts to learn Chinese, so that he could work at fragments from the Chinese herbal *Bencao gangmu*.[55] About the same time, Frederick Porter Smith, a missionary doctor in Hankou, compiled a treatise on Chinese *materia medica*, culling material from Chinese herbals as well as several Western works.[56] He embarked on this project mainly because he thought a better understanding of Chinese drugs might aid in the work of Western and native missionary doctors in China. The celebrated botanical collector Augustine Henry, a surgeon of the Chinese Customs, plunged himself into botany in the 1880s and 1890s partly as a recreation, but he was also powered by an interest in identifying Chinese medicinal plants and drugs.[57] He tried hard to decipher Chinese herbals and published several articles listing the identified items. Those who did not know Chinese found other ways to unlock the lore of Chinese *materia medica*. Charles Ford, Superintendent of the Hong Kong Botanic Gardens, developed an interest in Chinese drugs and collaborated with Ho Kai, a Chinese barrister with a medical degree from Edinburgh, on a project elucidating Chinese *materia medica*.[58]

Last but not least among the reasons that the naturalists sought information in Chinese works was their interest in geographical botany and the history of cultivated plants. The problem of geographical distribution of plants and animals had long been a major issue in natural history and had inspired different approaches and theories to explain the patterns, processes, and underlying causes of the distribution of floras and faunas.[59] The issue became even more important after the first decades of the nineteenth century, in part because of the influence of Humboldtian botanical geography. New and often controversial theories of migration, earth history, and Darwin's theory of evolution were also brought to bear on this challenging problem. Few sinologist-naturalists were equipped to enter this contested field, although many of them followed the discussion.[60] However, they did share the great Swiss botanist Alphonse de Candolle's goal of bringing human history into the study of botanical geography.[61] Candolle had included human agency in his list of means by which plants had been transmitted. If one wanted to find out how the current distribution of plants had come into being, human history had to be taken into account. Given their relevance to economic botany, agriculture, and the history of civilizations, Candolle argued, cultivated plants deserved special attention. He had also suggested that Chinese texts might offer useful clues to some of the questions regarding the history of cultivated plants.[62]

Emil Bretschneider, an eminent sinologist, wrote influential monographs to prove that Chinese works held keys to many questions regarding geographical botany and the history of cultivated plants. He argued that such works could be used to trace the origins of many cultivated plants and even to determine the routes of transmission.[63] For example, they provided information for deciding whether certain plants in China had been introduced from America or had been in the empire since time immemorial. Another example must suffice. Historical records held it that Zhang Qian (fl. 120 B.C.) brought back melons, grapes, and other curious fruits from his diplomatic mission to central Asia. While sinologists disagreed about the particulars of the story, they took it seriously and tested it against evidence collected from sources in other languages.[64] The sinologist-naturalists' intellectual background helped to explain why they were attracted to the particular areas of research. Their interest in Chinese civilization, their taste for literary studies, and their knowledge of sinology and philology drew them toward research involving these concerns. Bretschneider found systematic botany (namely classifying plants) "dried" and "monotonous," not the least because it left out "the relation of the plants to man."[65] He spoke for many of his colleagues in China.

Like natural history, historical geography and philology were popular subjects among Western scholars in China. The pages of the China journals were full of reports, queries, and heated debates on these subjects, which often dealt with the historical contact between China and other parts of the world, particularly India and central Asia. The cultural exchange among China, India, and central Asia had been substantial and left its traits in languages, religions, material culture, domesticated animals, and cultivated plants. To reconstruct that history from the elusive traces of evidence, sinologists in China applied orientalist scholarship, such as comparative philology developed by scholars of Indo-European languages, to their research into historical geography, ethnography, and natural history.

The sinologist-naturalists' interests in natural history and the other areas of research were not neatly compartmentalized. Not only did they use historical geography to inform their natural historical research, but they were able to mobilize natural history to assist their work in historical geography. For instance, Bretschneider, an eminent historical geographer among other things, and Sampson took part in a debate about the claim made by some European sinologists that a Chinese Buddhist monk discovered America in the sixth century. The monk's account included descriptions of interesting plants and animals he came across in the country. Sampson and Bretschneider culled evidence from Chinese botanical works to refute their European colleagues' assertion. The "Fu-sang" tree was not the Mexican aloe, they argued. And the country in question was probably Japan.[66]

Chinese Traditions of "Natural History"

Compared with the times of Old Canton, in the second half of the nineteenth century, the British in China had access to much larger numbers of Chinese texts. They could now buy books in major cities without difficulties. The increased availability broadened their contact with Chinese literature and enabled them to consult a much wider range of sources in their natural historical research. I have been very liberal in using terms like "Chinese works on natural history" or "Chinese botanical works," for the naturalists themselves used these descriptions. This habit of imposing their own categories on Chinese works originated, in part, from the general method of translating the unfamiliar by attributing to it the characteristics with which one was more familiar; that is, what has been called by Anthony Pagden "the principle of attachment."[67] It also reveals the scientific lens through which the sinologist-naturalists viewed the Chinese works. In fact, however, the Chinese did not have a discipline, a system of knowledge, or even a coherent scholarly tradition equivalent to Western notions of "natural history," "botany," or "zoology." The modern Chinese term for botany, *zhiwu xue*, the study of plants, was coined in 1858 to denote the Western science of plants. It first appeared in a translation of John Lindley's *Elements of Botany*. Similarly, the Chinese rendition of "natural history," *bowu xue*, was also a mid-nineteenth century neologism for translating Western works.[68]

On the other hand, the Chinese had produced a vast body of literature on animals, plants, and minerals, and they had developed various genres, methods, and intellectual traditions of analyzing and organizing them. For instance, the culture of gardens and gardening, so widespread among the gentry in the late Ming and Qing, promoted a vast gardening literature, including literary encyclopedias about ornamental plants, gardening handbooks, instruction manuals about flower painting, and so on. These writings were products of a literati that was busy breeding flowers, writing about them, and imbuing them with symbolism. They had books devoted to varieties of the orchid, the chrysanthemum, and the plum tree, classifying the varieties according to their aesthetic quality. Only within this "culture of flowers" can we explore what the literature meant to the Chinese audience.[69] The Western sinologist-naturalists were certainly aware that the Chinese works had been written for an audience with tastes and needs very different from their own. Yet they could also reorganize these works in scientific terms, thereby assessing the value of the works to their natural historical research. In the ocean of Chinese literature, what sources did the naturalists find most useful?

First, there were Chinese works that were devoted principally to plants, animals, and other natural objects. This literature shared enough similarities with the Western mode of natural history writing that Western naturalists readily likened it to their own tradition. However, Victorian naturalists in China, with Bentham's *Flora Hongkongensis*, Siebold's *Fauna Japonica*, Franchet's *Plantae Davidianae*, or David's *Les oiseaux de la Chine* side by side with a Chinese work, would likely find the latter a mess. To an unsympathetic Victorian naturalist, the Chinese way of ordering and describing the natural world could only be frustrating at first glance.

The most widely cited Chinese work was Li Shizhen's *Bencao gangmu* (1596), commonly translated as the "Great Materia Medica." It was mammoth in size and comprehensive in content. Filling thirty-six Chinese volumes (*juan*) in a seventeenth-century edition, it covered more than twelve hundred kinds of plants, animals, and minerals. It had been known to Western scholars since the early eighteenth century and remained "the text book for every student of Chinese natural history" in the nineteenth century.[70] Animals were grouped into five *bu*—insects/worms, scaly animals, animals with shells, birds, and beasts—each of which consisted of several *lei*. Birds, for example, comprised water birds, field birds, forest birds, and mountain birds. Each *lei* consisted of many *zhong*, which sometimes corresponded to species. Scaly animals were likewise organized into four *lei*, but the categories were based on appearance rather than the habitats of the animals: They were dragons (for example, the dragon, lizards, the pangolin), snakes, fish, and scaleless fish (for example, squid). Like animals, plants were grouped into five *bu*—herbs, cereals, vegetables, fruits, and trees—which were further divided into 31 *lei* and then innumerable *zhong*. Some of the subdivisions were defined according to habitat and others, by appearance and utility. So one *zhong* might correspond to one or several species or even items from widely different genera.

The taxonomy used in *Bencao* was derived from earlier works and, with variations, continued to be used in many later works of its kind. S. Wells Williams, an American missionary and friend of the botanist Asa Gray, thought the Chinese taxonomy "rude and unscientific" and, in referring to the botanical classification, declared that "the members of these families have no more relationship to each other than the heterogeneous family of an Egyptian slave dealer."[71] He placed *Bencao* "far behind the writings of Pliny and Dioscorides" and warned others that "its reputation tends greatly to perpetuate the errors."[72]

Not everyone shared his view. Otto F. Von Möllendorff, a zoologist and member of the German Consular Service, considered the system of zoological classification in *Bencao* "certainly equal, if not superior to, the systems of Ante-Linnaean Zoologists in Europe."[73] He also praised earlier Chinese works on natural history for the "signs of considerable acuteness of observation" exhibited therein.[74] Western zoologists in China focused mainly on mammals, birds, and fish—all of which were relatively big animals—and it did not require many observational details to figure out which animal is meant in a description. Consequently, they found the descriptions in some Chinese books reasonably accurate. Some Western naturalists also learned to be more appreciative of Chinese works as they became more familiar with them. As Joseph Needham noted, Bretschneider cast away most of his early criticisms of the Chinese works after eleven more years of study.[75] With sharpened perception of the similarities and differences between Chinese works and Western botanical literature, he recommended the Chinese works be "translated into European languages and . . . be commented upon."[76]

In *Bencao* and other substantive Chinese herbals, an animal or a plant is described, its synonyms listed, its medicinal qualities explained, and the places of production mentioned. The *mulan* tree may be taken as an example. Four synonyms were listed, each with the reference cited. The author then explained why the tree is called by these names. Then came descriptions, which included four quotations from earlier works and one from the author himself. They describe the shape and color of the leaf and flower, the size of a mature tree, its provinces of production, utilities, vernacular names, and so on. The author goes on to explain that the tree blooms in the fourth month of the Chinese calendar and that its blossoms fall in twenty days. It produces no fruits. He then comments on a couple of legends associated with the tree. Finally, he explains its medicinal qualities and usage, and gives prescriptions for some diseases; for example, the flower may be used to dissolve a fishbone stuck in the throat.

Bretschneider, and probably many others, did not find the writing in Chinese botanical works particularly difficult.[77] It was usually clear and straightforward. *Bencao* used the same organizing principles and descriptive vocabulary throughout. The color of the flower and the flowering season were always mentioned. Yet he still found "the descriptive details of plants" in the works "meager and unsatisfactory."[78] The Chinese did not have a shared set of botanical terminology, like botanical Latin in modern botany, that could represent the characteristic features of a plant in more or less standardized terms. In describing a plant, they often relied on comparing its organs with those of another plant. A European reader, who had usually never seen either of the plants, could hardly infer anything definite from the descriptions. Bretschneider lamented this problem, but acknowledged that the same technique was widely adopted in European botanical works until the time of Linnaeus.[79]

Verbal descriptions in Chinese works were often supplemented by illustrations.[80] *Bencao*, for example, included more than one thousand illustrations in wildly different styles, qualities, and origins. Like Renaissance and early modern European books on natural history, Chinese works often reproduced, quite indiscriminately, drawings from earlier books. Ernst Faber, a missionary and sinologist, complained of this Chinese practice that "the authors were commonly more familiar with books than with nature, and the block-cutters had no understanding of either."[81] Even today sinologists know practically nothing about how the Chinese used the illustrations in *Bencao*—many of which were simply too sketchy for the practical purpose of identification—and what these visual representations meant to Chinese readers. *Bencao* claimed that the drawings might be used as aids to identify the objects, yet, given the quality of the plates, there could only be a gap between the rhetoric and the practice in most instances. When the Chinese works included original drawings, delineated from nature, however, they could achieve a level of accuracy that impressed the naturalists. Bretschneider praised the woodcuts in *Jiuhuang bencao*, a treatise on plants for famine relief published in the early fifteenth century, claiming that many of them were "certainly superior to some European wood-cuts of the 17th century."[82]

Western botanists and zoologists held different opinions about the quality of the woodcuts in Chinese works on natural history. His praise of *Jiuhuang* notwithstanding, Bretschneider thought Chinese illustrations were usually "so rude that it is very seldom that any conclusion can be drawn from them."[83] Another botanist, referring to the plates in *Bencao*, declared that "it is often difficult to tell whether the figure is intended for that of a plant or a bird."[84] Albert Fauvel, on the other hand, used a drawing from the same work in his study of alligators in China.[85] Swinhoe and Möllendorff, both zoologists, found the pictures in a medieval edition of the ancient Chinese dictionary *Erya* useful.[86] Bretschneider, however, thought otherwise.[87] Some differences stemmed from the fact that more details were necessary in order to identify a plant as opposed to a big animal. Identification of the former usually required fine details of the flower and leaves, but these were often reduced to one or two strokes in a Chinese illustration. The only Chinese work widely admired for its illustrations was *Zhiwu mingshi tukao*.[88]

The author, Wu Qijun, did most of the drawings himself from live specimens. The figures were often exact enough for narrowing down the possibilities of the identity of a plant, so that further research would not be a wild goose chase. Bretschneider duplicated some of Wu Qijun's drawings in one of his influential treatises.[89] Sampson also included one in his paper on the *puti* tree.[90] When A. Henry tried to explain a plant to William Thiselton-Dyer, Director of Kew Gardens, he enclosed the figure of the plant from *Zhiwu mingshi tukao*.[91]

However unsatisfactory they might have appeared to a Victorian naturalist, *Bencao* and *Zhiwu* counted as the most empirical of Chinese works on natural history. The naturalists often had to wrestle with works whose purpose was primarily literary. One of the most widely used reference works for botanical research, *Guang qunfang pu* (1708), was commonly translated as the "Enlarged Botanical Thesaurus." In spite of this title, *Guang qunfang pu* was basically a literary anthology, composed of quotations from a wide range of Chinese literature. The quotations were arranged according to the plants. Although the naturalists seldom bothered to distinguish it from the genre of *bencao* or the herbal, lumping them together as botanical works, this work actually belonged to the tradition of gardening literature. In *Guang qunfang pu*, empirical observations mingled with poems, recipes, fables, prescriptions, and historical legends. The work was enormous: It comprised one hundred Chinese volumes and covered more than fifteen hundred kinds of plants. It was this comprehensiveness that made it a useful storehouse of information.

Large numbers of Chinese geographical works were also frequently consulted by Western naturalists in China. Not only did the naturalists' interest in historical geography and the history of cultivated plants lead them to the works, but fieldwork and the need to cross-examine accounts and references in an herbal prompted them to delve into Chinese geography. Chinese geographical writings encompassed various forms, but the gazetteer or local history was the staple source. The gazetteer, *fangzhi*, was an official publication, issued by the county, provincial, or prefectural government, that described in detail the history and geography of the jurisdiction. Westerners in China relied heavily on the gazetteers for intelligence about areas of which they had little firsthand knowledge. The gazetteer, as a rule, had a section on the natural products of the place, which listed and described local plants, animals, and minerals. The purpose of such lists was to showcase the natural riches of the area, and their quality varied greatly from one gazetteer to another.

Sketchy lists, like the one in the *Guizhou tongzhi* (general gazetteer of Guizhou, published in 1742), included a few hundred items. A majority of them had no annotations, just names. More detailed lists, like the one in the *Sichuan tongzhi* (general gazetteer of Sichuan, published in 1816), required dozens of Chinese pages, and the items mentioned were usually accompanied by brief descriptions. One of the most valuable features of the gazetteers was that they specified the place of production to the level of city or county. The naturalists could therefore direct their attention to a precise locality.

To be sure, these annotated lists hardly met the particular criteria of Western naturalists, whose strict demands for empirical accuracy and completeness would have baffled a regular Chinese local magistrate. Nevertheless the gazetteers provided Western naturalists with a handy starting point. In addition to information on natural objects, gazetteers also had sections on agriculture, manufacture, and trade, from which the naturalists could gather desultory but unique data on economic botany. When the British began to explore southwestern China in the 1870s and 1880s, they always took the local gazetteers with them on the road.[92]

The gazetteers could be important even to more specialized research. Although he complained about their rough quality, Swinhoe used gazetteers frequently in his pioneering studies of the fauna of Taiwan and of Hainan, an island south of the mainland, in the 1860s.[93] From them, he learned, among other things, about animals he had not previously known existed on the islands.[94] Möllendorff found a gazetteer "very useful" for studying the *vertebrata* in North China.[95] Bretschneider declared that the descriptions of natural productions in the gazetteer were sometimes "specific in great detail, accompanied by interesting remarks derived from local observations" and were of great interest to students of geographic botany.[96]

The sinologist-naturalists did not limit their attention to obvious sources, such as herbals and geographies. They consulted a wide range of other genres of writings, including travel literature, dictionaries, and even Buddhist texts, which contained information about the transmission of plants from India to China.[97] By the end of the nineteenth century, a large number of Chinese works had been combed through. Bretschneider's treatise on Chinese botanical works, *Botanicon sinicum* (1881), the most comprehensive work of its kind, listed more than one thousand references.

Textual Practice

The naturalist-sinologists in China were not the only naturalists who sought information in texts. At the beginning of the chapter, I proposed the concept of "textual practice," arguing that the use of texts was fundamentally important in nineteenth-century natural history. To further explain the concept, let's consider an obvious example—recently extinct species. When Victorian ornithologists were after the great auk, a newly extinct bird, they plowed through a wide range of works, including voyages, travelogues, geographies, and even old manuscripts kept in an Icelandic library.[98] Natural history was more than collecting fossils and mounting dead birds, doped butterflies, and dried plants. Nor was it limited to classifying specimens according to the Linnean or some other taxonomic system. Natural historians wanted to know the history of particular plants and animals, their habits and behavior, the nature and diversity of regional flora and fauna, and so on. These problems were crucial to the research of field naturalists and important to certain branches of natural history, such as economic botany and botanical geography, and they could not be answered by examining specimens or field trip observations alone.

In fact, even classifying a plant or animal required naturalists to check descriptions in old natural history books in order to avoid unnecessarily duplicating species and to keep descriptions simple and consistent. If a plant or animal had been properly described, it would not be acceptable to create a new species, much less a new genus. If old descriptions were applicable, it was advisable to adopt them as much as possible. Few naturalists in Europe or China had access to type specimens, even when such things existed, or any specimens of a given species at all; scientific descriptions in books usually were the only source for taxonomic practice.[99] Henry Hance advised a beginner to purchase a dozen or so botanical books, many of which were multivolume sets, for a start.[100] When E. C. Bowra, newly arrived to join the Chinese Customs, wanted to work on Chinese plants, the first thing he discovered was that he "must write home for some books on Botany" for he was "continually stumbling over" plants which were new to him.[101]

In a sense, therefore, consulting Chinese texts for natural historical research was not a radical departure from what a naturalist regularly did. He had to engage in textual practice in almost any research. Working at Chinese texts nonetheless presented unique challenges and problems. It required special training, knowledge, and scholarly techniques, and even a well-equipped researcher did not find his way easily in the labyrinth of Chinese literature. The works encompassed different genres of writing, spread over several centuries, and within a genre works were written in different styles. The language barrier had become less forbidding since the mid-nineteenth century, though the Europeans hardly considered learning Chinese easy. Bretschneider, one of the best sinologists of his time, repeated Ricci's statement from nearly three hundred years earlier: "for western people, Chinese is of all languages the most difficult."[102] But there was already a community of Western sinologists in China, and they could easily obtain linguistic assistance from native-speaking scholars. Basic linguistic difficulties were, therefore, no longer insurmountable. However, once sinologist-naturalists ventured into the literature on natural history, they could no longer depend upon their Chinese language instructors, "whose erudition seldom extends beyond the classics."[103]

Under this circumstance, familiarity with the literature itself became the surest guide. It was no easy task to track down particular plants and animals in Chinese works. The Chinese literary tradition of citing, quoting, and alluding to previous authors and works made the process less daunting, but the quest led immediately to philological detective work. To discover and locate the information, to comprehend the contents, to tease out the tangle of often-obscure names of people, plants, and places required a wide reading in Chinese literature.

Western naturalists also practiced certain interpretative strategies when approaching Chinese texts. Their tendency to compare the Chinese works with early Western herbals reflected a confidence in the superiority of their post-Linnean natural history and a disposition to arrange all cultural products on a ladder of progress. Both were characteristic of the comparative method of Victorians.[104] By drawing analogies between the Chinese works and early Western ones, the naturalists de facto ignored the fundamental issue of translation between two independent traditions of natural history. This attitude lay in sharp contrast, for instance, to the missionaries' toward Christianity and Chinese religions, which emphasized the categorical difference between Christianity and "heathenism" and which caused serious debates about correct translations of key concepts like "God."[105] More important, its uniformity differed from the diversity of opinions about the possibility of translating Western scientific works into Chinese and of teaching science in Chinese. Many of the Western intellectuals in China asked by the *Journal* of NCB in 1885 about this question thought that one could not teach or express scientific ideas in Chinese because of its linguistic limitations. They gave such reasons as its ideograms could not convey with full precision abstract ideas, logic, and rationality; it was extremely cumbersome and hard to learn even for Chinese themselves; and the innumerable neologisms required in order to translate scientific works accurately would make the whole enterprise unfeasible.[106] But others disagreed and actively participated in the missionary and the Chinese government enterprise of translating Western scientific works into Chinese.

When the translation worked the other way, from Chinese to English, the issue was no longer the intrinsic limitations of the receptor language. The sinologist-naturalists never thought twice about the cognitive and linguistic superiority of Western science and languages over Chinese ones.[107] They tacitly assumed that it was both possible and desirable to separate empirical facts in Chinese works from the muddle-headed Chinese system of knowledge. Most naturalists were not interested in exploring the general framework of Chinese lore on natural history. Many of them made casual remarks on the Chinese classification system of nature's objects, but none of them devoted more than incidental attention to it. I have found no serious attempt made to explain the conceptual foundation or the organizing principles of the Chinese lore. Western naturalists devoted little time to discussion about reconciling the two knowledge traditions or about absorbing the Chinese system *en bloc*. They focused instead on the practical matters of identifying individual plants and animals. The Chinese literature was treated more as a backward, immature, and miscellaneous source of information, like early Western works, than as the product of a tradition with its own history and logic. In their scientific research, the naturalists had first to "misinterpret" Chinese works or translate them to a new context of meanings, in which "a mass of other matter, appreciated only by Chinese readers" could be peeled off.[108]

The central task that the sinologist-naturalists set for themselves was to "fish pearls out of the mud": to sort out "useful data" from heaps of literary allusions, dubious anecdotes, and misinformation. However, even the mundane process of distilling natural historical data involved operations far more complicated than simply sieving out or selecting useful fragments. It involved textual and cognitive interpretation of Chinese works. There were no ready-made facts, labeled as such, to be collected. A piece of information, a description, a picture in a text became a fact only after it was interpreted to be one. Westerners had to bring their experience and training as naturalists and sinologists to the task. The hermeneutic process was embedded in, among other things, their understanding and perceptions of the Chinese literary tradition and of the Chinese people. They had to decide how much they wanted to trust the Chinese texts, and what to make of a particular statement in a particular Chinese text.[109] Personal experiences with the Chinese people, familiarity with Chinese literature, and a welter of other reasons might come into play in assessing evidence.

Whatever he based his opinions on, a naturalist's evaluations of Chinese achievement in natural history had direct bearing on the conclusion he drew from the literature. Swinhoe, for example, suggested in 1870 that the tapir had existed in China into historical time because its figure in an ancient Chinese book appeared to be delineated from nature.[110] Richard Owen, an eminent naturalist at the British Museum, however, was more skeptical: The figure could have depicted a tapir from Malaya, or drawn by a traveler to that region.[111] No powerful evidence could be cited to support either one of these possibilities. Swinhoe's hunch had come from his experience with Chinese works. Similarly, based on Mayers's study on maize in China, Hance was ready to conjecture that maize might have come to China from central Asia instead of America, because he believed that Chinese literature on natural history of the sixteenth and seventeenth centuries was in a "comparatively far more advanced state of development" than Western literature.[112] If the conclusion was derived from generally reliable Chinese books, he thought one should take it seriously.

The problem of interpreting and employing Chinese works also involved challenges and judgments of a more technical kind. For in-

Sinology and Natural History

stance, using the data gathered from *Guang qunfang pu*, a literary encyclopedia on plants, Sampson was able to put together a satisfactory (to him) description of the "*fung* tree." In the process, he rejected the statement of the fruit "being the size of a duck's egg" as "doubtless an oriental exaggeration."[113] Focusing on Sampson's cultural chauvinism or orientalism, however, would not get us far in explaining why he found the other descriptions from the same work trustworthy. More importantly, it would not tell us much about the naturalist's chief tasks in working with Chinese texts for natural historical research. This is not to deny that the discourse of natural history might contain an orientalist desire to collect the exotic and an imperialist will to name and order things from a Eurocentric perspective.[114] As we saw in Chapter 3, these are important issues. But to gain a deeper understanding of the research practice of the naturalists, we need to examine the nuts and bolts of their reading of Chinese works for natural historical research and to unpack the working of textual practice as a research method.

Inquiries into China's natural history often entailed a great deal of philological, geographical, and historical research, and it required the techniques and knowledge of the sinologist to parse, analyze, compare, and interpret the texts. The naturalists mobilized both natural history and sinology in interpreting Chinese texts, creating a discursive domain in which the boundary between the two disciplines was blurred and evidence from them valorized. Statements from a Chinese text might be taken as evidence to support or challenge arguments in Western natural history. Knowledge of natural history might be, in turn, cited to determine sinological queries.

From Earth Dragon to *Alligator Sinensis*

To illustrate how the sinologist-naturalists employed Chinese texts in their research, let's take a close look at a single example. Albert A. Fauvel's compact article on the Chinese alligator crystallizes the different aspects of the sinologist-naturalist's method and enterprise, and will serve our purpose well.[115] Although Fauvel was French, he worked closely with the British, as a member of the Chinese Customs and an ardent naturalist. He was in charge of the Natural History Museum of Shanghai, an establishment of the North China Branch of the Royal

Asiatic Society, in the late 1870s. His article on the alligator made a splash among Western scholars in China, in part because it was the first scientific paper to show the existence of alligators in China, but also because it advanced certain sinological inquiries. In fact, the article was as much a study in sinology as zoology, and its multifaceted and interconnected concerns reflected the scope of the sinologist-naturalists' research. Fauvel's interest in the subject had first been aroused by some reports of alligators in China. Accordingly he conducted a series of investigations, making field trips, examining actual specimens, and inquiring into Chinese literature. What he finally accomplished went far beyond a technical description of an alligator new to European scientists.

Fauvel's article began with a philological inquiry, trying to establish that certain Chinese characters represent the alligator (or the crocodile, for that matter), not any other animals that had been suggested by other Western authors. The character *tuo* (rendered as "Tó" in the paper), for example, had been previously identified as the iguana or lizard. Fauvel dismissed the first possibility based on his knowledge about the geographical distribution of the iguana, and, citing Chinese classics, he disputed the second possibility. The skin of *Tó* was used by ancient Chinese to make big drums, and no lizards in China could be large enough for the purpose. So *Tó* must be something else. He moved rapidly from natural historical to sinological arguments in the same paragraph and drew evidence from ancient Chinese texts as well as contemporary scientific works.

Fauvel devoted a significant portion of the article to philological discussion and explication of the synonyms that he had collected from Chinese classics, dictionaries, and natural history books. Although compiling and determining synonyms might appear to be little more than an after-dinner game for pedantic sinologists, it actually formed a cornerstone of the naturalists' research. For an empire with such a long history and vast territories, a plethora of synonyms for any plant or animal was only to be expected. Historical change and regional variation created confusion. The problem was further compounded by the coexistence of numerous dialects in the empire at any time in history, which gave rise to local names often wildly different from each other and from the common names. Alligators in China were a relatively easy case. Fauvel had to deal with only half a dozen archaic Chinese characters and terms. In the case of introduced plants, contorted foreign tones joined the babel of names. Plants introduced from India or central and western Asia could have arrived in China with phonetically translated names, which then went through a long evolution and many metamorphoses.[116] They could also have been named after places or after their shapes, tastes, colors, or smells. In an inquiry into Chinese figs, Sampson came up with more than a dozen plant names, probably all synonyms, whose origins he traced, respectively, to Sanskrit, Levantine, Persian, the appearance or taste of the fruit, and so on.[117] Not surprisingly, the sinologist-naturalists saw determining the synonyms as an essential step toward investigating a plant or animal. How else could one know which plant/animal was referred to in a statement in a Chinese work? Bretschneider suggested that his colleagues compile indexes of plant names, and Ernst Faber, a German missionary, heeded his call.[118] Augustine Henry provided himself with "the best Chinese books on plants," and commenced to identify the Chinese plant-names.[119] He confessed later that "such a work . . . can only be done very gradually" and that "[it] also requires an expert."[120]

Once the naturalist had the synonyms of the plant or animal in question, he could systematically collect the descriptions of it in Chinese works. In his paper, Fauvel turned to Chinese works on natural history, including *Bencao gangmu*, to see what they had to say about *Tó*. *Bencao* placed the animal in the category of dragons, and provided synonyms and descriptions. The animal was said to be an earth dragon and it was described as looking like a gigantic gecko or pangolin. It was extremely strong, slept a lot, cried at night, and laid "up to one hundred" eggs. Its back and tail were covered with armor. Fauvel quoted at length the descriptions from *Bencao* to demonstrate that, with one or two exceptions, they agree "wonderfully well" with those by Audubon and others of the alligators of the Mississippi and Guyana.[121] The "somewhat quaint but sufficiently characteristic picture" in *Bencao* was reproduced to back up his argument.[122]

Having shown that *Tó* was the alligator and having collected the synonyms, Fauvel then tackled the issues of history, distribution, and the alligator in Chinese culture and society. Fauvel had two excellent libraries to work at—the one at the NCB and an even better one at the Jesuit missionary station at Xujiahui, only a few miles from Shanghai. He found next to nothing about alligators in works by early Western travelers and missionaries in China. As I have argued, grubbing about in old travelogues, geographies, and so on for data remained a strong tradition in nineteenth-century natural history, and Fauvel was operating within the tradition. While early Western accounts proved to be of little value, luckily Fauvel had at his disposal numerous Chinese works. Citing as authorities Chinese classics, herbals, dictionaries, encyclopedias, local histories, and so on, Fauvel argued that historical records bear out that alligators used to rule the lakes and rivers in central and southern China, though they had disappeared almost entirely in recent times. Testimonies in Chinese books were absolutely critical to Fauvel's reconstruction. In China there was no possibility for Westerners to do systematic paleontological fieldwork of the kind they might pursue in Europe. They had to be content with historical documents. While naturalists in Europe were busy reassembling the remains of prehistoric animals, Mayers dug out of ancient Chinese books descriptions and drawings of a huge, hairy, rat-like creature living underground, which he believed to be the mammoth.[123]

Fauvel's sinological interest led him to explore the alligator in Chinese legends and mythology, thereby examining the cultural meanings of the animal in Chinese history. This section of the article addresses the concerns of sinologists. The last part of this article is a scientific description based on the specimens sent to him, which he believed to be a new species. Only with specimens could he determine whether the alligator was a new species, as descriptions in Chinese books were too crude for taxonomic exactitude. But even this part of his research depended heavily on books. Without access to the specimens of alligators from other parts of the world, Fauvel worked from taxonomic reference books. His discovery of a new species proved to be uncontroversial and was immediately confirmed by the Muséum d'Histoire Naturelle.

In fewer than forty pages, Fauvel not only presented a zoological discovery but he also made some sinological excursions. He determined the meanings of certain Chinese characters and elucidated a few aspects of alligators in Chinese life and literature. Natural history and sinology were deeply interrelated in his examination of the history of alligators in China. With textual evidence, he had reconstructed the habitats, behaviors, and vicissitudes of alligators in China.

Natural History and Knowledge Translation

In this chapter, I have mapped out the intersections between sinology and natural history in the nineteenth century in relation to the Western presence in China. Western diplomatic and missionary institutions in China expanded rapidly during the second half of the nineteenth century. This development fostered the growth of sinology and provided talent and infrastructure for natural historical research. A group of scholars whose interests encompassed natural history and sinology found a forum for their research in the *JNCB* and other Western journals published in China. Influenced by their humanist background, they directed their attention to economic botany, geographical botany, and the history of cultivated plants; their philological training provided them with powerful research tools. The tradition of textual practice in natural history allowed them to incorporate their sinological learning into their natural history research.

To naturalists in Europe, the sinologist-naturalists' work certainly appeared distinctive, involving as it did the detailed reading and interpretation of recondite sources in a forbidding language. Although few scientific journals in Europe were willing to publish articles detailing such specialized philological and historical inquiries, European scientific communities nonetheless took great interest in the conclusions that were to be drawn from the sinologists' research. Their ignorance of sinology did not prevent them from appreciating the efforts of their colleagues in China. Alphonse de Candolle's influential *Origin of Cultivated Plants*, for example, relied heavily on Bretschneider's treatise on Chinese botanical works.[124] Bretschneider, Sampson, and others had identified numerous Chinese plants and used Chinese texts to determine whether they were exotic, naturalized, or indigenous species. The information contributed to the mapping of global biogeography.

My research has revealed that textual scholarship was more relevant to nineteenth-century natural history than has been assumed. In fact, Darwin himself used many examples from Chinese texts to support his arguments in natural history. Attempting to prove his hypothesis of natural selection, which he compared to breeding and artificial selection, Darwin endeavored to compile evidence about the domestication of animals and plants in history. In 1855 he asserted that many abnormal varieties of domestic animals had originated from China and Indo-China, "the China-men evincing a most especial taste for monsters of all kinds (witness their carvings), & shewing the same perverted taste in their floriculture, dwarfs of trees, &c&c."[125] Losing no time in pursuing this topic, he contacted potential correspondents in many parts of the world, but especially in China, for specimens of "any domestic breed or race, of Poultry, Pigeons, Rabbits, Cats, & even dogs," which had been "bred for many generations in any little visited region."[126]

Domestication and breeding had been going on for centuries; how could one reconstruct the history of domestic animals and cultivated plants? Specimens would of course be valuable, but it was necessarily difficult to assemble physical evidence from different historical periods or to trace the past from contemporary specimens. Therefore, Darwin sought additional evidence in old Chinese texts, asking a curator at the British Museum for help. By this means, he obtained translated accounts of the breeds of Chinese fowl from the *Bencao gangmu* and the encyclopedia *Sancai tuhui* (1609). Later, Darwin also received sinological assistance from his correspondents in China. Prompted by Darwin's remarks on gold-fish in his *Variation of Animals and Plants under Domestication* (1869), Mayers researched the topic and published a paper on gold-fish cultivation in Chinese history in *Notes and Queries* in the same year. Hance, a friend of Mayers, immediately forwarded the paper to Darwin, who then incorporated it into his *Descent of Man* (1871) and the revised edition of *Variation* (1875).[127]

Finally, my study shows that the transmission of knowledge across cultures, which inevitably involved translation and interpretation, raises important questions about the intersecting realms of knowledge, power, politics, and cultural traditions. The sinologist-naturalists practiced cross-cultural translation of knowledge. They translated Chinese knowledge about animals and plants into a new context of meanings. Thus, an earth dragon whose flesh was of medicinal value became *Alligator sinensis*, which was distinguished from other crocodilians by its dentition and other anatomic details. Legends and histories were turned into evidence to construct the geographical distribution of the animal in the past. In making their translations, the naturalists assumed that the cognitive and linguistic superiority of Western science and languages empowered them to dissect Chinese knowledge and to select the factual data from this knowledge. Identifying this imperialist conception alone, however, is not enough for a detailed analysis of the sinologist-naturalists' research into Chinese works. When they encountered particular Chinese texts for their scientific investigations, they constantly had to draw boundaries between facts and false information contained in those texts, but the process was informed by many factors that cannot be reduced to mere imperialist ideology, the European gaze, worldview, or any other broad epistemological units. A nuanced understanding of the scientific discourse in which the sinologist-naturalists combined textual evidence, philological data, historical documents, and indigenous knowledge of the living world with Western concepts of natural history can only be attained if one follows the historical actors themselves and observes how and why they made the decisions that they did during their encounters with Chinese texts.

5

Travel and Fieldwork in the Interior

PART OF THE attraction of natural history came from its appeal to the senses and sentiments, its engagement with bodily, aesthetic, religious, and emotional experience. To its Western practitioners, natural history meant much more than merely a rational, analytical exercise of intellect or a practice of dogged empiricism. To reduce the enterprise of natural history to theoretical debates, institutional networks, and economic utilities, therefore, is to render the history of natural history dry and colorless. Visual experience, physical exertions, and the joy of discovery were all important components of practicing natural history. The aesthetic and sentimental dimension of time spent strolling in the woods, watching birds fly, or discovering a rare plant was woven into the naturalist's experience of studying natural history.[1]

Perhaps no other activity in natural history was more filled with intense action and experience than fieldwork, especially fieldwork conducted in an unfamiliar land.[2] Whether high in the Himalayas or deep in the Amazon jungles, Victorian naturalists' expeditions, often full of trials and dangers, demanded psychological fortitude and physical stamina—a situation at which the naturalists never tired of hinting in their travel accounts. The traveling naturalists valued the experiences of an expedition as much as the specimens shipped home.[3] They did not venture into the forests and mountains simply for the specimens and scientific data. The process and the experience mattered a great deal. They recorded and interpreted their experiences in diaries, letters, and travel accounts. The process of travel was thus imbued with meanings, meanings that dialogized with allegories of civilization and the persona of the Victorian explorer-naturalist. There were heroes and villains, forces human and natural, events surprising and predictable, anecdotes comic and tragic. There were encounters between bearers of the torch of civilization and savages, primitives, barbarians, or conceited Orientals, for, however remote from "civilization," Western scientific travelers were rarely by themselves. The mysterious, the unknown, the space devoid of Western presence might actually be some other people's backyard. What was a heroic adventure to the Westerners might be an everyday routine to others.

In the second half of the nineteenth century, Western naturalists and scientific travelers explored a significant portion of the Qing Empire. The Russian explorer Nikolai Przhewalski crossed Mongolia and Xinjiang (East Turkestan); the French missionaries Évariste Huc and Armand David roamed China proper, Mongolia, and Tibet; the German geologist Ferdinand von Richthofen visited most of the eighteen provinces; and British naturalists and travelers, including Robert Fortune and Robert Swinhoe, traveled in China proper, Manchuria, and the islands in the China Sea.[4] The access of Westerners in China to the vast interior increased as the century unfolded, and the naturalists seized the new opportunities for conducting fieldwork in the areas hitherto closed to them. The modes and circumstances of travel in China differed greatly from similar adventures in some other parts of the world, and these differences directly affected the practice and experience of the naturalists pursuing scientific research.

Not only did China's flora and fauna vary significantly from north to south, east to west, but China had a population of four hundred million, made up of peoples with distinct languages, life styles, and histories.[5] Qing China was going through societal upheavals and natural disasters with breathtaking frequency and on a stupendous scale. Revolts were frequent. The country was also faced with external pressures on several fronts. It had been warring with foreign imperial powers from time to time since 1839. Its people held diverse opinions about Westerners and Western things. Foreigners traveling, hunting, or doing fieldwork in China inevitably confronted these situations, and the naturalists were highly aware of them. They never worked in an insulated bubble, protected from the social realities and cultural conflicts, but constantly had to adjust their actions to the local circumstances and to negotiate with different groups of Chinese.

Most of the British naturalists' expeditions were limited to China proper, which had been well trodden long before any Europeans set foot on its land. The Chinese were not trees: they moved; they traveled; they had trade and commerce, roads and transport; they drew maps; they wrote and read travel accounts; they visited sites of historic interest; they had a long tradition of pilgrimage; they had their famous mountains and favorite resorts; they had inns, taverns, and tea houses everywhere and brothels for travelers here and there.[6] Exploration in China was, therefore, nothing like David Livingstone's expeditions in Africa or Henry Walter Bates's on the Amazon.[7] Western naturalists in China took full advantage of the Chinese infrastructure of travel. More often than not, they traveled by chair and passenger boat or, in northern China, by cart or horse.[8] They consulted native maps, using them to correct Western ones.[9] They usually moved from village to village, from town to town, rather than struggling through treacherous swamps and damp rainforests. At night, they rested at an inn and dined well.[10] When they entered a large town or a city, which dotted the routes, the local mandarins received them as important guests. This relative comfort, however, did not prevent the travelers and naturalists from producing adventure narratives of conventional drama and turns of plot. Such accounts were saturated with exoticism, geographical imagination, imperialism, and the ideology of masculinity represented by confrontations with nature, although there were also accounts emphasizing leisure and pleasure rather than physical and psychological hardships.[11]

Travel and fieldwork in natural history cannot be properly understood without taking into account the location, the indigenous people, their folk knowledge, and the power relations between traveling naturalists and the natives participating in the fieldwork. In this chapter, I shall first explain the aims and modes of the naturalists' travel and fieldwork in China. Because their activities almost always involved the Chinese, it is necessary then to examine the Chinese guides, collectors, and other informants working with the naturalists and to unravel the fieldwork relations among all the participants, including the local people. To carry out their field research, the naturalists mined Chinese folk knowledge, so part of this chapter discusses the role of this lore in fieldwork and natural history in general.

Scientific Explorations in China

It is possible to group British natural historical explorations in China into major categories according to their purposes and modes of research. As we saw in Chapter 3, many British residents in China were engaged in collecting botanical and zoological specimens. They were usually bound by official duties and collected whenever opportunities arose. But there were also British scientific travelers, either professional collectors or gentleman naturalists, visiting China for the purpose of collecting natural history specimens. Several ships constantly surveyed the seas and coasts of China, designated as the China Station by the British Navy.[12] Like many of their brethren, notably Thomas Huxley, a few naval surgeons and naturalists voyaging in the region devoted their attention to marine biology.[13] Arthur Adams published numerous papers on mollusks in the Japanese and Chinese seas during the middle of the century.[14] In the same manner, Cuthbert Collingwood, a surgeon-naturalist who visited Taiwan and the China coast in 1866–1867, and later P. W. Bassett-Smith collected diligently in the area.[15] Naval surveys barely touched upon the coasts, so the shipboard naturalists had but brief contacts with China. Except for comparative purposes, therefore, I do not include this type of exploration in the following discussion.

Botanical prospecting and collecting, the second type of exploration in view, figured prominently in the history of British research on China's natural history and can be traced back to the times of Old Canton when Kew and the Horticultural Society of London first sent collectors to China. After the Opium War, plant hunters ventured into the interior and eventually reached as far as the Sichuan, Yunnan, and the Tibet border, where they found a paradise of botanical riches.[16] A third kind of expedition was mostly associated with zoological research and increased its activity only after the mid-nineteenth century. The Victorian culture of sport and travel, popular throughout the British Empire, had its local variety among the British in China.[17] Hunting and zoology often went hand in hand, as many field naturalists were ardent sportsmen.

The three different kinds of scientific exploration—coastal survey, plant hunting, and field zoology—hardly exhausted the range of British scientific explorations in China, but they were representative. Geological expeditions, which might seem to some fieldwork *par excellence*, were rare in China because political circumstances limited the opportunities to excavate and conduct other extensive fieldwork.[18] Research had to rely largely on impressionistic observations, as did the studies by Raphael Pompelly, an American geologist, and Richthofen, unless one includes the surveys for developing telegraph and railroads.[19] The curator of the Shanghai Museum confessed in 1878 that "Chinese geology . . . is little known [and] Chinese fossils are not known at all."[20] Organized scientific expeditions carried out by teams of naturalists, notably Roy Chapman Andrews's expeditions to the Gobi, did not make their appearance in China until the early twentieth century. In the rest of the chapter, my discussion follows closely the nature of British scientific expeditions in China, focusing on botanical collecting and zoological exploration. It must be kept in mind, though, that most scientific travelers collected a variety of natural history specimens with an emphasis on one or two particular areas of research according to their interests, knowledge, and missions.

Kew Collectors

Ever since the times of Old Canton, British naturalists in China had wanted to explore the southeastern interior, where tea was produced for export, and the lower Yangzi region, where garden plants suitable for temperate climates grew. Immediately after the first Opium War, the Horticultural Society of London sent Robert Fortune to China for the express purpose of collecting new garden plants.[21] Fortune would visit China three more times, including two expeditions into the tea districts to bring tea plants and tea manufacturers to India to found a tea plantation there. He brought back numerous new ornamental plants, clarified many questions about planting and manufacturing tea, and earned popular fame as a China traveler through the success of his travel books.[22] The opening of Japan in the 1850s further induced the British to send plant hunters to East Asia. The island nation, whose latitude was similar to that of Britain, was thought to possess garden plants, trees, and so on that would adapt well to British soils.

Kew Gardens sent two collectors, Charles Wilford and Richard Oldham, to China and Japan in the 1850s.[23] Private nursery firms quickly followed suit, sending Fortune, John Veitch, Charles Maries, and others.[24] In the last decades of the century, British plant hunting in China entered another phase because the plant hunters now could reach western and southwestern China proper, previously accessible only to French missionaries. Augustine Henry botanized in the area in the 1880s and 1890s, thereby paving the road for professional plant hunters, notably Ernest Henry Wilson, nicknamed "Chinese Wilson," and George Forrest, who introduced many new rhododendrons to Britain.[25]

Plant hunting offered young gardeners and naturalists a great opportunity for "seeing the world." It excited the romantic imagination of traveling among strange people in exotic landscapes. It also gave them an opportunity of carving out a more successful and exciting career than would otherwise have been possible. Fortune's phenomenal success must have inspired many young men to follow in his footsteps. Plant hunters sent to China by the Horticultural Society of London and Kew Gardens were poorly paid. At a mere £100 a year, the collectors had to pinch every penny, and they repeatedly protested about it.[26] The salary seemed adequate for a gardener in England, but things were so dear in the Western communities in China that the collectors could barely keep themselves sufficiently fed and dressed with that amount. Because almost everything they collected belonged to the institutions, they could scarcely make profits from the collections.[27]

Traditionally, if the mission of a plant hunter for Kew or the Horticultural Society was successful, yielding many important discoveries, his future might be bright. He would "have acquired a distinction & *status*, which [he] could hardly attain in any other way."[28] His experience as a plant hunter could secure him a good position in a major private nursery. Or he might be appointed director of a colonial garden, as was William Kerr, who, after his service in China, was transferred to Ceylon to supervise its botanical gardens.[29] Hoping to be promoted in this way after his three-year expedition in China, Oldham was bitterly disappointed.[30] Penniless and frustrated, he severed his ties with Kew and planned to follow Fortune's career path, collecting for private nurseries, but he died only a few months later.[31] The career of Wilford, Oldham's predecessor, was not so much tragic as exasperating. He simply refused to carry out his mission, resolutely being irresponsible, and got into many financial troubles, embarrassing Kew Gardens as much as himself.[32]

Why did the two Kew collectors struggle in their missions while Fortune, the collector for the Horticultural Society, accomplished so much? Personal qualities, no doubt, made a difference. Fortune evidently had all the right stuff for a plant hunter to China—education, independence, energy, fortitude, street smarts, and, not the least, an amiable nature—"a fine stout, healthy-looking man," according to Robert Hart, a keen judge of character.[33] By contrast, Wilford left his hosts in China, including Henry Hance, with a poor impression.[34] Oldham, however, seemed to be a fine man and hit it off with Robert Swinhoe.[35] What prevented him from doing more? The answer was clear to Hance and Oldham himself.[36] Kew Gardens had assigned the collectors to naval vessels. William Hooker's idea was that this arrangement saved expenses and provided mobility. The collectors could travel on the survey ships to many places far away, increasing efficiency and covering more territory. But the Kew collectors ended up spending months at a single port, and when they were on survey voyages, the ships hardly paused at a place long enough for them to collect on shore.

Hance and Oldham both cited the example of Carl Maximowicz, who was then collecting in Japan for the Botanical Garden of St. Petersburg, to show Hooker how botanical collecting should be done in China and Japan. The key, they argued, was to pick a reliable man and give him the freedom to judge where to go and how to proceed in his expedition according to the changing circumstances. He should also have sufficient funds to employ native collectors. While Maximowicz had teams of Japanese collectors working for him, Oldham had few and was soon forced to tighten his purse strings.[37] Although Fortune, too, was running on a stingy budget, he at least had enough control of his itinerary in China. He traveled far from the urban centers, whereas the Kew collectors were either languishing in port or confined to their ships. No wonder they felt that their hands were tied. At the end of his mission for Kew Gardens, Oldham sincerely wished in a moving letter to Hooker that no botanical collector would ever again be sent to China and Japan under the same conditions. Luckily, as we shall see, the Kew collectors' experience was not typical of British traveling naturalists in China, whose mode of field research more or less resembled those of Fortune and Maximowicz.

Travel, Hunting, and Natural History

Victorian gentlemen in China shared with their countrymen elsewhere interests in sport, travel, and natural history. The three activities often went together in the British imperial context. With the expansion of the British Empire in the nineteenth century, globe-trotting became an industry: Missionaries, traders, colonial officials, and (later) tourists busily sailed among Europe, the colonies, and other parts of the world. In China, the British developed a local culture of travel and tourism. They traveled extensively along the Yangzi River and among the treaty ports. At certain times of the year—summer vacation, Christmas holidays, Chinese New Year, for example—the British combined hunting trips with leisure travel.[38] By the end of the century, they had developed a culture of travel whose favorite destinations included the Three Gorges, Beijing, Canton, and the ancient cities in the lower Yangzi provinces.[39] They also found in Chefoo in north China an attractive summer resort, akin to the hill stations in India's Darjeeling district, and large numbers of Western residents in Shanghai flocked there on vacation to escape the heat.[40]

In these aspects, the British lifestyle in China closely resembled that of their compatriots in India. A welter of tourist guides, travel literature, and sportsmen's handbooks appeared on the market form the 1870s onward. The trend of travel in China reached far among the British around the globe. Newspapers, including the popular *Illustrated London News*, sent special correspondents to China to do travel reports.[41] The restless Isabella Bird, one of the most famous Victorian women travelers, sailed up the Yangzi River.[42] Henry E. M. James and Francis Younghusband ventured into Manchuria from India on an expedition to pursue big-game hunting as well as to explore the area.[43]

Because China produced few strikingly unusual mammals or birds, it did not excite the European imagination of exotic fauna as did Australia or Africa. Deer, wild pigs, wolves, and tigers were among the most impressive animals that China had to offer. In Victorian imperial culture, big-game hunting enjoyed high prestige because of its display of valor and skill—hence the safari in Africa and the tiger hunt in India. Although there were fewer opportunities for similar adventures in densely populated China proper, hunting was extremely popular among the British in China. Everyone hunted.[44] The settled areas ravaged by the Taiping Rebellion reverted into wildernesses roamed by pheasants, deer, and wild pigs. These places became favorite hunting grounds of Western residents in the lower Yangzi region.[45] However, birds, rather than mammals, were the primary victims of Western sportsmen in China. Shooting pheasants, while not exactly a heroic endeavor, had its charm.[46] Pheasants, snipes, partridges, quails, and waterfowl were so numerous in the Yangzi region that a marksman could bag hundreds of birds in a few days. When size was not an issue, numbers mattered.[47]

In the last quarter of the nineteenth century, sportsmen sailed in houseboats up the Yangzi River, sometimes all the way to Yichang, or along the many smaller rivers in different directions to new locales for bird and deer shooting.[48] The scale of this sport was such that many sportsmen themselves attributed the rapid decrease of game birds in the popular hunting grounds to unrestrained shooting and pleaded for conservation measures. They took the lesson from their fellow sportsmen in Britain and the colonies, though others were convinced that the Chinese hunters, whose numbers increased greatly in response to the market demand of Shanghai and export, and the expanding agriculture in the areas caused more destruction than did the Western sportsmen.[49]

Although most British sportsmen in China would not consider themselves naturalists, their numbers included several serious students of zoology and ornithology, for example, Swinhoe, who was active in the 1860s, and later F. W. Styan, C. B. Rickett, and J. D. D. La Touche.[50] Many took the trouble to collect specimens for their naturalist friends in China and for the Shanghai Museum, founded in 1874. The Museum was overwhelmed with birds given by sportsmen from all the treaty ports. Admittedly, the sportsmen's interest did not always correspond to the naturalists', and donations tended to be of the same birds, which were either obtained at the treaty ports or in the popular hunting grounds, hardly representative of the species richness and geographic distribution of so vast a country as China.[51] When regular sportsmen went hunting, they paid attention only to certain desirable birds and mammals. As a result, some groups of animals were disproportionately represented at the Museum. Insects, reptiles, and small mammals like mice were ignored.[52] Bona fide naturalists like Swinhoe and La Touche, on the other hand, collected systematically in the field with the aim of studying the fauna of a particular area. Nevertheless, the widespread interest in sport, travel, and natural history among British residents in China supported the serious zoological research of the more dedicated, who shared their passion for hunting game.

Modes of Collecting

Long-distance, continuous travel was not the typical mode of fieldwork research for Western residents in China. Both French and British resident-naturalists shared a regular mode of research; that is, systematically conducting fieldwork in good collecting grounds. Typical field trips of theirs involved short trips between the residence and a few collecting grounds within the radius of a few days' trip. This approach could be as productive as extensive traveling. Augustine Henry, for example, made only two long field trips, which together lasted a few months, during the whole of his career. His extraordinary success was built upon thorough collecting in the places to which he was posted. Rickett and La Touche exhausted the avifauna of the Fuzhou area.[53] If a location was rich in unknown flora and fauna, then a comprehensive search would likely turn up abundant novelties. Henry's Yichang Gorges, Swinhoe's Taiwan, and the French missionary Jean M. Delavay's western Yunnan testified to the effectiveness of this approach. This mode of research made a virtue of necessity, for, burdened with professional duties, the resident-naturalists had no opportunity to travel for months or years on end. The only exception was long holidays.

By comparison, Fortune, the entomological collector Antwerp E. Pratt, and the plant hunter Wilson traveled constantly for the purpose of collecting specimens, though their ranges still paled beside those of French missionary-naturalists, like Armand David and Pierre Heude.[54] (In fact, British travelers in remote parts of China often depended on the hospitality of local French missionaries, for the French had a wider network in the deep interior than did the British.) Traveling naturalists did not usually stay at a place long. They passed through a series of places that might or might not welcome strangers. Confronted with language barriers and physical vulnerability, they often needed reliable natives to help them carry out their projects. On his second journey to China, in 1848, Robert Fortune had his head shaved, wore a fake queue, and donned a Chinese robe. "I made a pretty good Chinaman," he reckoned.[55] But he still needed two "authentic" Chinese as guides, porters, collectors, and interpreters. They turned out to be, in addition to all these roles, "troublemakers," and poor Fortune constantly worried that they would abandon him on the road. Luckily, they didn't, and the trip ended fruitfully.

The Chinese who accompanied Western naturalists in travel and fieldwork usually included a body servant (called a "boy") who also usually acted as an interpreter, a cook, a local guide, several collectors, and a dozen or more porters who carried the luggage, expedition supplies, and carefully packed specimens. Guides and porters could be easily obtained through Chinese agencies that provided services to Chinese travelers all over the country. This method of travel was commonly adopted by Western travelers in China.[56] There were also professional Chinese interpreters and guides specializing in Western clientele. In Hong Kong and Canton, where Western tourists abounded, several professional tour guides were available, including Ah-Cum, who showed a young journalist, Rudyard Kipling, around Canton in 1889.[57]

For his journey to the Tibetan border, Thomas T. Cooper engaged an interpreter and guide who had switched his career from the Catholic priesthood to commerce and who spoke Latin, English, and probably some other Western languages, in addition to his mother tongue, Chinese.[58] The French missionary-traveler Huc had a Catholic Mongolian who would later lead David from Beijing through Mongolia to Tibet. And when the American geologist Pompelly, who had read Huc's books while a boy, visited northern China in 1863–1864, he was delighted to discover this man and immediately enlisted his service.[59] The collectors in fieldwork included the naturalists' household servants, contracted collectors, and those hired on the spot during the journey. In his journey to Sikkim by way of India, Joseph Hooker employed scores of native collectors, guides, and porters.[60] Few British naturalists in China attempted such large-scale endeavors, but Pratt and Wilson, both professional collectors, did mount expeditions approaching this scale. At any rate, the basic categories of fieldwork personnel remained the same, variation in size of retinue notwithstanding.

In his many wanderings in China, Fortune engaged a small entourage, depending largely on collectors hired on the spot. He found children particularly useful for this end, because a minimal amount of money would "[go] a long way with the little urchins."[61] It was a practice widely adopted by traveling naturalists. On his voyage about Taiwan in 1866, Collingwood hired children to gather insects and shells for him whenever he found a chance.[62] It saved labor and time, and made good use of the unique knowledge and skills of the children.

Benefiting from this practice, Arthur Adams, a naturalist well traveled on the Chinese seas and coasts, thought that children were his best allies. More than anyone else, children knew the hiding places of beetles, frogs, reptiles, and other hard-to-find creatures, and they were ruthless in their pursuit of the spoils. Tree bark was peeled; earth turned; cow dung broken and examined. In China, these village children were everywhere, ready to work (and *play*) hard for a little cash. Dozens of them flew about the fields, in the woods, and by the ponds, and brought back all sorts of surprises and beauties. "They showed as much ardour in the chase as any naturalist."[63]

Of course, the traveling naturalists also employed adult collectors on the spot; and unlike the arrangement with children, this association could last a season or longer. A visiting naturalist could hardly find better collectors than the villagers, who knew the habits and habitats of local animals and plants better than any Westerners did. For the more labor-intensive tasks such as collecting insects, troops of natives paid cheap wages proved to be highly efficient. Most collecting required only minimal training in techniques, and, with the help of his guides and interpreters, a naturalist could usually recruit many local peasants to collect in the field. Some of them were even willing to follow the naturalists' expedition until their service was no longer needed, as long as they found the pay and travel worthwhile.

An experienced entomological collector, Pratt took full advantage of this method of collecting. Traveling up the Yangzi from Yichang, he identified the most promising collecting grounds and arranged for the natives to collect for him. He taught them the basics of collecting and preserving insects, handed out butterfly nets, assigned foremen, and then proceeded in this fashion into western China. At certain spots, he might hire as many as twenty or thirty collectors. On his way back at season's end, he either revisited the places to pick up packed specimens or had the collections brought to him. He invariably found the results highly satisfactory.[64]

Fortune had used the same technique on his plant-hunting trips. He made the island of Chusan, still occupied by the British forces at the time, his headquarters, and from there he made several forays into the tea districts of Zhejiang and Fujian. In Zhejiang, he stayed at the monasteries in the mountains and collected specimens and information about processing tea. In each place, he arranged with the natives to collect tea seeds for the season and, once the arrangement was made, immediately moved on to another place where he repeated the same process. Instead of traveling aimlessly, Fortune focused his collecting efforts in specific collecting grounds where he engaged the locals to collect plant specimens. He visited these preferred sites so many times during his journeys in China that he became a routine guest of the local monasteries.[65]

This technique of collecting was particularly suitable for traveling professional collectors, who strove to acquire the largest collections possible during a short period of time. Efficiency was the ruling principle. A professional collector was always busy, constantly on the road to the next site. He worked from morning to evening, racing against the season. To maximize the rewards of the season, a collector could not rely on his energies alone. The best method was to make an arrangement with the natives before the season and leave the actual collecting to them. In this way, the collector could reap the full season's fruits.

The Chinese had their own traditions of collecting plants and animals, but usually for purposes other than natural history in its narrower sense. In their fieldwork, the naturalists learned that they could not entirely disregard the Chinese traditions. Upon arriving at a village one morning, Fortune told the children there to catch insects for him. When he returned later that day, he was astonished to see that a crowd of peasants, old and young, men and women, were waiting to sell him insects, most of which, to Fortune's chagrin, were incomplete and entomologically useless. The Chinese had mistaken him for the traditional collector in China, a medical man who collected insects to make drugs. The creatures would end up in mortars, anyway, the Chinese reasoned, so what difference would it make whether they were whole or broken?[66] Theophilus Sampson, who collected in various parts of Guangdong, discovered two tricks for enlisting the villagers' help in collecting specimens. He either told them that he was an herb doctor or appealed to "the Chinese sense of beauty, and if [they] be persuaded that I am gathering plants really because they are pretty, I rise in their estimation as a man of refined taste."[67]

Hunting and collecting herbs were family traditions in China pursued by mountain dwellers or other country folk. Some naturalists managed to establish business relationships with these rustic Chinese; or perhaps the Chinese sought out a business relationship that promised economic advantage. Swinhoe employed "a vast number of native hunters and stuffers" when he was stationed in Taiwan.[68] The naturalists also encouraged the local Chinese—farmers, fishermen, or whatever their occupations might be—to bring them any animals that were interesting or unusual. Many naturalists even employed native collectors on a long-term basis, sometimes lasting many years. The naturalists entering this kind of partnership were, of course, not traveling naturalists, but residents in China, whose rigid schedules did not allow them to make long expeditions. When Westerners took holidays, usually in summer, the flowering season of many plants was already over. Employing native collectors was the best possible solution to this problem. "From my own experience," Bretschneider testified in 1880, "the best way to procure good herbarium specimens is to entrust an intelligent Chinaman with the commission."[69] Several zoologists, too, found it necessary to hire native "shooters" for a long term, for collecting birds and mammals required time-consuming expeditions into the countryside and mountains. Styan, La Touche, and Rickett all employed long-term collectors.

This kind of operation could function only when the collectors and the naturalists developed a reliable working relationship that was founded on trust, expertise, and economic interest—basic elements of the moral economy of fieldwork. The naturalists first considered whether the collector was a "conscientious" and "trustworthy" man. Henry, who had employed many long-term collectors, relied on Chinese Catholics and recommended them to others.[70] They had already had much contact with and respect for Westerners, which reduced the risk of misunderstanding and disagreements. Their religious background also reassured British naturalists about their honesty and reliability, for Chinese and Catholic though they were, they were at least not treacherous heathens. They could be employed through Western Catholic missionaries. Presumably, Chinese Protestant converts also fit the criteria, but they tended to be an urban and suburban lot, lacking the knowledge and skills to conduct fieldwork in the countryside. According to Henry's experience, the ideal arrangement was to hire Catholic converts to collect in their botanically rich home districts.[71]

The Chinese who indiscriminately gathered all sorts of specimens and data was not a good collector. To fulfill his assignments, for example, a native zoological collector needed to know not only the basics of collecting and taxidermy, but also how to note the habits and habitats of the animals he collected. Rickett's collectors, two brothers, worked for him from the 1880s to the 1890s. In addition to collecting in the Fuzhou area, they regularly visited the Wuyi Mountains in western Fujian to buy specimens from the hunters there and collected themselves along the way. The two brothers conducted fieldwork in western Fujian throughout the season. Rickett entrusted the whole matter to them; in fact, he had himself never visited some of the collecting grounds deep in the mountains.[72] Also in these decades, Styan's shooter traveled from Shanghai up the Yangzi to Hubei and Sichuan, more than 1,500 miles from the coast, to collect birds and mammals there.[73] While in Yichang, Henry employed the same collectors for years; they usually collected in their home districts, but Henry occasionally sent them to areas near the Yangzi Gorges on trips that could last months. Obviously they had the ability to operate independently.[74] The specimens they acquired this way were so impressive that Kew gave Henry a grant to fund further such expeditions.[75] When Henry was stationed on the Yunnan-Laos border, he had a collector he praised highly. The collector soon took on full responsibility for all the collecting, as Henry and his collector had exhausted the area surrounding his post and had to resort to long-distance field trips to yield novelties.[76]

These contracted collectors, mostly peasants and hunters, probably left their homes in rural or mountainous areas to sell their labor, and were hired to collect in their homelands. The botanist Bretschneider had a "simple peasant living in the mountains west of Peking" collect for him.[77] These people did not have profitable vocations, nor did they enjoy much social esteem; however, their familiarity with the local flora, fauna, geography, and language proved to be an invaluable resource to Western naturalists. In addition, labor was comparatively cheap in China—a fact that did not escape the naturalists and that helps to explain why they considered employing native collectors practicable even when, more often than not, they had to pay wages out of their own pockets. While stationed in Taiwan in the early 1890s, Henry paid his collector about eight to ten silver dollars (approximately £1.5) a month for two gatherings of "dried or semi-dried plants, of 50 or 60 species each time, each species being represented by 4 or 5 sheets of specimens."[78] He paid his collectors in Yichang in the 1880s less.[79]

Indeed, it cost little to carry out a botanical expedition in the interior of China. According to Henry's calculation in 1899, two thousand silver dollars a year would support a sizable team consisting of a "boy," six coolies to carry baggage, ten coolies to collect and carry collections, and maybe some mules, all expenses included—food, lodging, and everything.[80] The hunters employed for collecting animals must have received comparatively higher pay for their exertions because of their expertise, the difficulty of the task, the ammunition, and the higher market value of the trophies and specimens. In his letters to the British Museum, Styan made it clear that he expected the museum to pay adequate amounts for the specimens he sent them because he "employed several natives to hunt regularly for [him]" and the expenses incurred in collecting were "heavy."[81]

The skills and knowledge that the collectors acquired in fieldwork rarely had any application outside the work relationship with the Western naturalists. One notable exception was David's collector Wang Shu-hang, who became the taxidermist at the Shanghai Museum, which attracted hundreds of Western and Chinese visitors a year.[82] The position would later go to La Touche's collector, Tang Wang-wang, who was, according to his employer, "very capable" and "an excellent field naturalist."[83] Several other members of the Tang family would also collect for Western customers, and "field collecting became almost a clan profession."[84] But most native collectors probably returned to their previous occupations as peasants, hunters, or casual laborers. Collecting for Westerners did not interrupt their lives too much because they had always worked in the field anyway. They had made only slight adjustments to meet the requirements of their Western employers and had no difficulty whatsoever returning to their former lives. In a sense, Western naturalists and Chinese collectors alike exploited the opportunity of fieldwork and collecting: The naturalists searched for fruitful results, and the Chinese worked for the extra income they could not have earned through their own trades. But this static functionalist analysis sheds little light on the actual process and dynamics of cross-cultural relations that to a great extent determined the outcome of fieldwork. The issue of fieldwork relations warrants a closer look.

Power, Performance, and Cultural Encounter

Natural history research in the field required constant negotiations between the naturalists and the natives they employed and between the naturalists, as Western travelers, and the local population. In China, the relationships between the local population and the guides, interpreters, and collectors working for Westerners also played out in complex terms. Fieldwork is important to our understanding of scientific imperialism, not simply because it was crucial to natural historical research, but also because it was a site of power struggle and provides a window on scientific imperialism in action at the level of micropolitics in everyday life.

It is important to remember that the distribution of power in fieldwork relations did not necessarily favor the naturalists.[85] Even in the colonies, European naturalists in the field often found their authority undermined by the natives' subtle tactics of resistance. The naturalists were highly dependent and vulnerable in unfamiliar places. A native guide or porter could steal cash or supplies, hinder the expedition's progress by stalling or "soldiering," run away, willfully give misinformation, or sabotage the equipment or specimens; any of the tactics could have damaging effects on the naturalists' mission. The naturalists often found themselves at the mercy of their native guides—not to mention weather, diseases, poor roads, and other complications. In China, their control over such circumstances was even more tenuous.

However much troubled by foreign aggression, China was never colonized by Western powers, and British naturalists on expeditions in China did not enjoy the direct support and protection of the British imperial apparatus. Westerners who traveled beyond the treaty ports needed to petition the Chinese government for permission. The passports would specify the places the holders could visit and the conditions under which they could travel.[86] If a Western traveler violated the specifications, local governments could detain him or refuse him passage. The explorer T. T. Cooper, for example, was placed under house arrest by a local official—although it is also true that Chinese authorities tended to appease Westerners in these kinds of situations in order to avoid diplomatic controversy.[87]

On an expedition, the Chinese collectors saw themselves first of all as casual laborers working for cash. If they felt they were underpaid, they bargained with their employers for higher pay or protested. But they also valued personal and social relations, and expected to be treated properly as they understood it. Besides money and social skills, the naturalists did not have any effective means to rein in their Chinese employees, who, when unsatisfied, might quit their jobs at any moment. While collecting in western China in 1890, one of Pratt's most trusted men ignored his duty and sneaked out to seek pleasure in a city.[88] On another expedition, a Chinese collector brought Pratt a venomous snake that Pratt had wanted; and when this specimen was rejected, the collector "threw it among [Pratt's] children."[89] But many performed their work conscientiously. Pratt was impressed with one of his collectors, who worked very hard despite suffering a nasty snakebite.[90] The plant hunter E. H. Wilson employed "about a dozen peasants," whom he found "[f]aithful, intelligent, reliable, cheerful under adverse circumstances, and always willing to give their best."[91] When they parted at the end of the expedition, Wilson felt that "it was with genuine regret on both sides."[92] A good personal and working relationship between naturalist and Chinese employee was vital to successful fieldwork. When consulted by Charles Sargent, Director of the Arnold Arboretum, about the necessary qualities for a plant hunter in China, Henry highlighted, besides an enthusiasm for travel and botany, "commonsense, tact and especially good temper."[93]

Travel and hunting in China inevitably led to contact with the Chinese, and the biggest trouble possible to Western travelers and sportsmen was not any natural obstacles or wild beasts, but conflicts with the natives. Books on travel and shooting in China typically contained many tips, compiled from the collective experience of Western travelers and sportsmen in China, on how to handle difficult situations. There were always villagers working in the fields, hidden from sight amidst crops or reeds, and shooting accidents could occur. Hunting handbooks advised the readers how to assuage the anger of the natives should this happen and suggested a proper compensation to the injured. If the situation turned threatening, the sportsmen were advised to seek protection from the local authorities.[94]

In the second half of the nineteenth century, many parts of China barely recognized the authority of the Qing government. In much of southwestern China, whose flora and fauna were particularly interesting, neither Western powers nor the Qing government could guarantee the complete safety of Western travelers. In 1875 in Yunnan, Augustus Margary of the Consular Service was mysteriously killed while surveying the road from Burma to western China, and the incident almost sparked a diplomatic crisis.[95] For fear of more such incidents, the local authorities usually assigned guards and runners to accompany Western travelers. Even in places where the Qing government had a firm grip, the safety of Westerners was occasionally threatened. Mobs attacked missionaries in Hunan, Sichuan, Guangxi, and elsewhere in the last decades of the nineteenth century.[96] Meanwhile in northern China, similar incidents, compounded with political struggle within the Qing government, culminated in the Boxer Uprising in 1900.[97]

Passing through angry towns, Western travelers might be hooted at and pelted by the crowds; the American bicyclist Thomas Stevens had this bad experience while visiting China on his journey around the world.[98] Charles Maries, a collector for Veitch's nursery, sailed up the Yangzi to Yichang in 1879, but retreated after only one week because of local hostility.[99] He thus missed a golden opportunity; only a few years later, Augustine Henry would astound the botanical world with his numerous discoveries from the same area.

The entomological collector Antwerp Pratt worked for a while in a part of Hubei notoriously hostile to Westerners. Notices in Chinese were "pasted on the trees all around," warning that "any native who was found to be collecting for [Pratt], or assisting [him] in any way," would be punished.[100] Later, on the Chinese-Tibetan border, he found himself haunted by the same problem. No locals would work for him except the Catholic converts, whom he employed through the help of the French missionaries. Rumors circulated about Pratt's "true" intentions in coming to the district, and anger and fear were in the air. His assistant, a German collector, was driven from a collecting ground by the local Chinese. Even his Chinese collectors encountered similar difficulties. The local people harassed them and blocked their fieldwork.[101] When young George Forrest went to western Yunnan to collect for the Edinburgh Botanic Gardens in 1905, he ran into a local uprising against the French missionaries and their Chinese converts. The missionaries were killed; Forrest, "being hunted like a mad dog," escaped only after nine days of hiding and fleeing barefoot through the hills.[102]

However, curiosity rather than fear or hatred was the more usual reaction of a Chinese local population toward Western strangers. British travelers typically described how, when entering a town, they were trailed by large crowds of local people, gazed at, and commented upon. Their every move excited murmurs or laughter. Trapped in this circumstance, Swinhoe compared himself unfavorably to an aye-aye scrutinized by a group of naturalists.[103] Other Western travelers also likened themselves to persecuted animals. A group of British explorers rowed into a town in the inland province of Guangxi one evening and woke up next morning to see "all the streets and boats near [them] were packed with people anxious to get a sight of foreigners." The explorers "tried to escape them by . . . darting hither and thither like a bird trying to escape from a hawk," but to no avail. At their wits' end, they took turns exhibiting themselves on their boat to the crowds gathering on the riverbank.[104]

Under this relentless Chinese gaze, the Western travelers strove to live up to the image by which they defined themselves vis-à-vis Orientals, and they acted out the ideology of imperialism in the play of symbols. The working of cultural performances of imperialism in everyday life is well illustrated by an instance highly relevant to travel in China. The sedan chair borne by bearers was a common transport in much of central and southern China, and Westerners traveling in these areas routinely used it. Although a white man, stiffly attired, sitting high on a chair carried by barefoot Chinese evokes for us everything appalling about Western imperialism, this particular mode of transport had existed in China long before Westerners came. It was widely used by Chinese travelers who could afford it and thus differed from the system of *silleros*, Indian carriers, in the Andes, which had been developed to take white people across the steep mountains.[105] Indeed Western customers were extreme rarities compared with the Chinese who daily patronized the chair as transportation. But Westerners in China imbued symbolic meaning into the image of "a white man riding the Orientals," hence the numerous stereotypical pictures of this motif produced in the era.[106] A transport representing the customers' social status in the Chinese context was translated into racial superiority and imperial domination.[107]

Meanwhile, the idea of muscular imperialism and a passion for hiking in nature often induced Victorian travelers in the countryside to walk much of the way, surely to the puzzlement of their Chinese chair-bearers, and to mount the chairs only when they were entering towns and cities, in accordance with the manners of the Chinese gentry. They of course did not forget to mention in their travel accounts how they walked along ragged mountain trails, rather than being carried (by implication, effeminately) on the chair. Referring to an exploration in southwestern China, E. C. Baber, a consular officer, claimed that he and his fellow travelers "never used our chairs except when required to do so by etiquette." They surveyed and hunted all the way, and even had a Chinese taxidermist with them.[108] Alexander Hosie, consul and correspondent of Kew Gardens, went on a journey to investigate the secret of the valuable insect white wax. He came to the celebrated Mount Omei, began climbing it, and found himself among a stream of pilgrims. The climbing became exceedingly difficult; but driven by "that British pluck," he finished it without resorting to the carriers.[109]

These cultural and racial performances sometimes even cornered the Westerners into the irony of imperialism or colonialism brilliantly evoked in George Orwell's "Shooting an Elephant." In the story, a colonial officer in British Burma is forced to shoot an elephant "solely to avoid looking a fool" before the natives.[110] Swinhoe once got himself into such a predicament; while traveling, he was requested by some Chinese to shoot a tiger that had caused havoc in the village. Lured by the imperial sportsman's ultimate trophy, the glossy skin of a tiger, and driven by his English sense of pride, he took great risk to shoot the beast with only a fowling piece while the villagers, without useful weapons, watched from a safe distance and ignored Swinhoe's shouts to chase down the wounded and furious animal. The tiger escaped; Swinhoe had almost gotten himself killed and was bitterly disappointed.[111]

The imperialism embodied in these subtle cultural performances pervaded everyday encounters between naturalists and Chinese.[112] Often the Westerners felt compelled to act in ways that demonstrated to themselves (and hopefully the Chinese) their superiority. Of course, these cultural performances might misfire or simply appear eccentric to the eyes of the Chinese, who had no ideas about the cultural values represented by the Westerners' actions. On a journey to southwestern China, Hosie scaled a snow-covered peak while his Chinese clerk and servant were "well mounted" on chairs. Mystified, the Chinese witnesses on the road held a discussion on why the Westerner preferred to walk when he could have been carried.[113] There were no guidelines, and British scientific travelers just had to act according to their best judgment in different circumstances. In his exploration in Manchuria, Younghusband displayed his scientific equipment prominently, a tactic "calculated to impress the unsophisticated natives."[114] On the other hand, T. T. Cooper was advised by an old China hand not to bring any odd-looking scientific equipment on his expedition to western China, lest it provoke Chinese suspicion.[115] At any rate, British imperialism in China was hardly limited to commercial aggression backed up by gunboat diplomacy. It also permeated cultural practices in everyday life. Technological superiority, particular codes of honor, cultural confidence and arrogance, a desire to export universal values, and a belief in "handling the natives firmly" all entered into the thinking of British travelers and naturalists in the field and shaped their conduct toward the affairs concerning the Chinese.

Folk Knowledge

Another important but easily neglected aspect of naturalists' activities was the collection of folk knowledge. It was as fundamental a component of fieldwork as collecting specimens and making field observations. Current literature on botanical and zoological collecting has focused almost exclusively on the collection of specimens and presents a very incomplete picture of fieldwork. It is true that nineteenth-century naturalists did not usually rely on field observations and folk knowledge to classify animals and plants according to established systematics; this was mostly done back at the museum. Yet fieldwork data frequently contributed to the practice of classification.

For example, Robert Fortune's observations in the tea districts in China proved crucial to settling the long-running debate on the species of different kinds of tea. European botanists had long been intrigued by the question of whether black tea and green tea were the same or different species. Like Linnaeus, many regarded them as products of different plants. Export black and green tea came mainly from two separate provinces of China to which the Westerners had no direct access, and the tea plants reputedly from the two places seemed to many to be of different species. Others disagreed and classified them as varieties. Processed tea leaves were of little use in this matter. Fragments of information gathered in Canton and the few tea plants of uncertain origin introduced to Europe since the seventeenth century only added fuel to the controversy.

In his journeys to the tea districts, Fortune discovered, to his surprise, that the Chinese made black and green tea from leaves of the same plant, an observation confirmed by his Chinese informants. Thus different kinds of tea leaves were simply the products of different methods of processing. The tea trees in the "black" and "green" tea provinces indeed differed somewhat from each other; but, based on his extensive observations, Fortune concluded that they should be considered as cultivated varieties rather than separate species. The success and high publicity of Fortune's mission must have added authority to his conclusions, and his testimony effectively ended the controversy, despite a few dissenting voices.[116]

Indeed, fieldwork and folk knowledge even seeped into discussions on major theoretical issues, as evinced in the writings of Darwin himself.[117] At a time when the concept of species attracted much attention and when the habits of the creatures were often discussed within this very context, fieldwork observations seemed to have some theoretical weight in this matter. Swinhoe once speculated that the habits of animals (birds, in this case) should be counted as a category of classification of species, in addition to physical characteristics.[118] Influenced by Darwin, Swinhoe was experimenting with the concept of species. He noted that despite their close physical resemblance, certain sparrows in China had developed very different habits from those in Britain. Should they therefore be classified as two different species? And, in another instance, he ventured to remark that, "If difference of note is to be taken as a guarantee of difference of species, then must we consider the many Eastern forms [of the cuckoos] distinct."[119] Admittedly, Swinhoe did not take these speculations too seriously; for whatever theoretical merits his idea might have had, it would have posed incredible difficulties for the practice of classification. The elusive nature and inherent ambiguity of the "habits" of animals necessarily made Swinhoe's speculations highly unpromising. As a field naturalist, Swinhoe probably attached more importance to the insights gained from fieldwork than did most museum naturalists.

Nevertheless, both fieldwork and folk knowledge bore direct relevance to many areas of research in natural history, including "hot" issues of the day, like geographical distribution and migration. The most reliable way to put together a picture of the distribution of certain plants and animals was based on fieldwork data. Only fieldwork could determine the location where particular animals and plants were found. Like his correspondent Alfred Russel Wallace, Swinhoe was interested in the geographical distribution of species, particularly across South and East Asia. Drawing upon his extensive field experience in China, he proposed that the Himalayan type of fauna stretched all the way to Taiwan. His fieldwork on the island of Hainan also suggested to him that the separation of the island from mainland China must have been relatively recent, as its fauna very much resembled that on the mainland.[120] Similarly, the cumulative discoveries of British and French botanists in the deep interior of China eventually supported Asa Gray's hypothesis of intercontinental phytogeographical distribution of trees across East Asia and North America, made famous by his battle with Louis Agassiz, another giant of nineteenth-century American science, over Darwinism.[121]

In the field, the naturalists also actively collected folklore on local plants and animals. To take a familiar example, they always took great interest in the natives' lore on the utility of certain products of nature. As we have seen in the previous chapters, the British tapped the practical knowledge of Chinese gardeners, collected information about Chinese *materia medica*, and inquired about natural products of economic value, such as rice paper, insect white wax, and the tallow tree. Most of the information had been gathered from common folk. Transplanting or acclimatizing a plant involved a whole package of practical knowledge about the soil, environment, and care the plant needed to survive and flourish. On his tea-hunting trips, Fortune gathered some of these data by direct field observations, but most of his information came from his Chinese informants. Without their help, he could not even tell exceptional tea plants from merely ordinary ones.[122]

No one was more familiar with the local plants and animals than the people who lived among them and observed them daily. Farmers, fishermen, and hunters were in close contact with the local flora and fauna, and they knew a great deal about them. Their work and daily life required them to make careful observations of many of the creatures. Although the knowledge thus accumulated was not written down, it was extensive and contained unique information. Nineteenth-century geologists often drew upon the experience of miners and quarrymen. In their investigation of the great auk, a bird of contemporary extinction, the British ornithologists Alfred Newton and John Wolley visited fishing villages in Iceland and interviewed nearly one hundred fishermen for eyewitness accounts and other information.[123] Although few field naturalists gathered folklore so systematically, the underlying assumption—that is, a recognition of the value of folk knowledge—was widely shared by Victorian naturalists.

Traveling naturalists rarely stayed in one place long enough to become fully familiar with its flora and fauna. Most likely, they would not be able to observe the seasonal succession of plants and animals or many important habits of the animals, or note when and how the fur or plumage was shed. Robert Fortune was at times forced to gather ornamental plants whose flowers he had not seen because the season was already over. Despite many months of diligently studying birds in Taiwan in 1861–1862, Swinhoe admitted that "[a] great deal yet remains to be learnt of the habits of particular species."[124] Often naturalists had to rely on local informants to supply details. The information they sought from the natives ranged widely from the most basic questions, such as where to look for the plants and animals they wanted, to the more complex, such as the diet of a particular animal or the medicinal properties of a given plant. Does the plant flower? When does it flower? What is it called? An ornithologist might try to obtain from the natives information about the migration pattern of a certain bird, how many eggs it laid, and other such particulars.[125] Specimens, whether dried plants or bird skins, provided no answers to these kinds of questions, and even direct observations made by a trained field naturalist during a brief stay often proved insufficient.

In his travels throughout central and western China, Armand David constantly inquired about the local animals and those he might expect to encounter further along his route.[126] The knowledge helped him focus on the most interesting animals, so that he could efficiently use the limited time at his disposal. David also noted down the natives' descriptions of the habits and diet of the animals he procured. He learned, for example, that the jumping mouse was nocturnal, herbivorous, and lived in deep burrows, and that the yellow rat was diurnal and lived, several individuals in a group, in shallow dens.[127] Without this kind of help from the natives, a traveling naturalist could not hope to discover rare or reclusive animals and would have no clue where to look for them. As a traveler, he rarely had the opportunity to make consistent field observations. In fact, many of the animals were brought to him by his collectors or the local people, and he never saw them in their natural state; most he never saw alive.[128] This experience was common among traveling naturalists in foreign lands. Swinhoe wrote about a bird that "[this] specimen was brought to me . . . by my hunters. . . . I never met with it alive."[129] While in Borneo, Wallace received from a Chinese settler a tree frog with long toes and large webs between the toes. He never saw the frog in its natural habitat, but accepted the Chinese informant's description that the frog could fly or glide with the huge webs expanded.[130]

The data supplied by the natives did not automatically gain the naturalists' credence, for although most were straightforward descriptions—the diet, habits, history, varieties, and distribution of certain birds, for example—the process of evaluating them could be complex. Cunning, as much as jealousy and conceit, was seen by Westerners as one of the worst and most universal vices of the Chinese, although naturalists did feel that, compared with officials, petty shopkeepers, and town folk, Chinese farmers, mountain villagers, and other country folk in the interior were less deceitful.[131] They also commented that the Chinese sometimes agreed simply out of politeness or an eagerness to please. A naturalist therefore warned others never to "lead the witness" and put questions to the Chinese in a suggestive way.[132] As mentioned in Chapter 3, moreover, Western naturalists concluded that the Chinese were extremely credulous and did not care about the truth. They were careless about giving their opinions. When asked to identify a plant, they would tell their foreign interlocutors "the first name that comes into their head."[133] Then, there was the belief widely shared by Western naturalists that the Chinese were incorrigibly unobservant. They did not know the names of the trees under which they walked every day.[134] They would say that a tree did not flower even though it flowered every season right before their eyes.[135]

In representing the Chinese in this light, Western naturalists at the same time defined the qualities of a good Westerner and a good naturalist—he was one who valued honesty, facts, and truth.[136] He was the opposite of the Chinaman. Yet we should not allow these images of the Other to detract our attention from the working criteria the naturalists adopted when interpreting Chinese folk knowledge. Their penchant to reduce the Chinese to the discourse of Otherness was challenged by real-life experiences; and despite all their reflexive slighting comments, they usually gave the data gathered from their Chinese informants serious consideration.

Some naturalists devised and conducted a systematic and elaborate screening process to test the reliability of their Chinese informants. In 1878, assisted by his colleague Henry Hance, Edward H. Parker, a member of the Consular Service in Canton, gathered more than two hundred local plants and asked "at least three natives, entirely unconnected with and widely separated from each other," to identify each of them. "At least fifty natives," including some herbalists, were thus invited to name the plants. According to the rules of the test, the opinion of one herbalist weighed as much as those of two "rustics," most of whom were probably coolies, servants, gardeners, and farmers. Parker also consulted Chinese herbals; each author's statement equaled the opinion of a rustic. One of the herbalists named "with great readiness over one hundred dried plants," but only a few of his identifications matched those taken from the rustics and books. Parker naturally wanted to test how reliable this herbalist's authority was, a curiosity intensified by the herbalist's refusal to name some exotic plants, mixed with the local plants, given to him. The "native Galen" was invited to go through and name the same plants again ten days later, but "failed miserably." Citing this example, Parker warned his colleagues that "[the] evidence of Chinese herbalists should be used with caution."[137]

The general results of the whole experiment seemed disappointing. Parker concluded that the Chinese paid "little attention" to these plants.[138] The natives often applied the same name to two or more plants, that is, two or more distinct plants under the Linnean system. The naturalists were hardly surprised by "sloppy lumping" of this kind in folk taxonomy because of their firm belief in the superiority of their own classification system. Swinhoe once coolly reported that the Chinese of a particular city did not distinguish between two different species of deer, and, on another occasion, he ridiculed the Chinese country folk for their crude classification of birds.[139] Similarly, Parker noted the promiscuity of vernacular Chinese names. Lack of consensus on the classification of plants was hardly a problem unique to the Chinese, however, and Parker conceded that Western botanists also disagreed among themselves about some of the plants on the lists: Were they distinct species or simply garden varieties? At any rate, this elaborate experiment confirmed the naturalists' generally low opinion about Chinese folklore on natural history, at least as far as classification was concerned.

Who, then, were the "most competent Chinese authorities" that could be found in the treaty ports? Gardeners and herbalists, many naturalists agreed. Their occupations and expertise made them the most reliable sources of information. But they did not necessarily know a great deal about the ordinary wild plants; some of them probably had never ventured out of the city limits.[140] After surveying druggists in Chengdu, Sichuan, Alexander Hosie, H. M. Consul, declared, "They [the druggists] know not, and do not care to know, whence the raw materials are derived or whence they come."[141] The criticism would probably have puzzled his informants, whose job was, to use Western terminology, pharmacy, not botany. Despite all their complaints, the naturalists still sought information from these native experts. In fact, after their investigation, Parker and his botanical friends in Canton decided to train "an old native doctor" and thought that he would be of "considerable use as a corroborator."[142]

The real experts on wild plants and animals were hunters, herbal collectors, and peasants who worked in the districts remote from the towns and cities. When the naturalists pursued fieldwork outside the familiar urban and suburban setting, they naturally had more opportunities to meet people whose living depended on wild plants and animals. After working through a collecting ground in the mountains in Fujian, La Touche, himself an experienced ornithologist, praised the five hunters in a local village: "As field naturalists the hunters are really No. 1." "Few birds . . . are unknown to them."[143] The only exceptions were small warblers and some other tiny birds of no interest to the hunters. Moreover, he added, "[A]ll their statements regarding the natural history of the district were perfectly truthful and straightforward."[144] Unlike their urban counterparts, Chinese in remote villages appeared simple, guileless, and reliable in the naturalists' eyes—an impression derived from the collective experience of Westerners in China and colored by idyllic idealization. Fortune, who repeatedly accused the Chinese of untrustworthiness, tempered much of his faultfinding when he spoke of the villagers and farmers he encountered in the tea districts.[145]

La Touche's remark gives one an impression that naturalists tested the natives' reports against their own field observations whenever possible. That might indeed have been the norm; in some cases, however, the naturalists dismissed outright the statements of their Chinese informants, not because they had any direct counterevidence, but because the statements appeared incompatible with the established tenets of Western natural history. Rickett was greatly amused by some notions held by the local shooters and fishermen, who informed him that snipes were bred in the mud of the rice fields, that certain birds were produced from floating jellyfish, and that there was a bird with only one leg.[146] Rickett and his collectors had relied heavily on these people for specimens and information. These hunters and fishermen observed the local birds every day, whereas Rickett, a bank manager, stayed in his office from nine to four, six days a week. Yet Rickett did not rush out to investigate these supposed marvels.[147] He simply assumed that the Chinese must be wrong. Their descriptions violated certain concepts so fundamental to natural history, so firmly rooted in the consensus of the scientific community, that they were dismissed offhand as vulgar errors.

In most cases, naturalists tried to make sense of the raw information provided by the Chinese informants. A native once described to David the migratory birds annually passing through the country where he hunted. He could distinguish several species of cranes and three species of swans. The peasant was a Catholic convert and an experienced hunter, so he fulfilled the basic criteria of a good informant—personal quality and expertise. He probably had also been supplying David with birds and had earned his confidence. David had no difficulty in trusting the hunter's words, but he found it necessary to reinterpret the data. According to the hunter, one of the three swans was gray, and David concluded that it was "undoubtedly only the young of one of the others."[148] Here David resorted to plausibility reasoning.[149] The hunter's acquaintance with the birds was evidently limited to the brief encounters during the migratory season, and he set the young birds apart from the other two birds because of their distinct color.

Although the naturalists necessarily exerted their authority in determining what might be true and what was worth reporting to the scientific community, thereby policing the boundaries of science, they were frequently forced to leave the matter largely unresolved. They might highlight the sources of information, like "I was told by the natives," "I have it on the authority of the natives," "I was informed by my hunters," and so on.[150] Such references admitted that the naturalists simply could not decide one way or another and that they had to stop short of

Travel and Fieldwork in the Interior 151

completing the boundary work of inclusion/exclusion. Or as happened often, the naturalists employed the plausibility criteria commonly accepted in the scientific community to distill the data. In either case, a borderland straddling natural history and folk knowledge was opened up. As collectors and interpreters of folk knowledge, the naturalists appropriated the lore of data, but were not able to fully discipline it. Their expertise in natural history could carry them only so far, and the information gathered from the common folk, in different degrees of scientific plausibility, kept the border porous and ambiguous. Fieldwork was a central part of nineteenth-century natural history. Much of the knowledge in natural history was produced in the field, in a foreign land, through negotiations with the natives, and the accompanying influx of folk knowledge, in myriad forms, into all areas of natural history suggests the need for a broad reassessment of the knowledge formation of natural history in the period. This objective is of course beyond the scope of the present investigation.

Narrating, Mapping, and Imperial Space

The British had been exploring China since the 1840s and, by the end of the nineteenth century, they had reached the frontiers of the Qing Empire. Traveling westward across China proper, by land and boat, they entered mountain-ringed Sichuan, whence they ascended the Tibetan plateau, hiking along trails covered with mud and snow, and approached Lhasa. They explored southwestern China on horseback, on foot, and on the shoulders of Chinese coolies. They wound through miles after miles of poppy fields in Sichuan, turned south to plague-ridden Yunnan and famine-stricken Guizhou, and followed caravan routes into Burma and Siam. They left footprints on the summit of the Changbai Mountains in Manchuria. They crossed the Gobi Desert in Mongolia. The British were wrestling with the French for control of Indo-China, and the Great Game with Russia ranged wide across central and eastern Asia. There were talks about the imminent collapse of the Qing Empire, about the partition of China. Exploration in China cannot be separated from this growing presence and intensity of Western imperialism in China.[151]

Travel meant movement. It also entailed inscription—narrating, mapping, sketching, chiseling names on the trees and rocks, leaving doggerels on the walls of taverns, and, perhaps less frequently, erecting milestones and monuments.[152] British traveling naturalists in China "lived through" the travels. They *experienced* them. They also interpreted their experience as travelers, sportsmen, and field naturalists. Many of them wrote about their travel and fieldwork experiences—in letters, articles, or books. Narrating one's travel experience had become a routine in the great age of the Victorian exploration. Robert Fortune's travel books were enormously popular, went through several printings, and made him the most famous British traveler in the China of his time. His humor, entertaining anecdotes, and vivid descriptions certainly helped sell books. But the very fact that the site of his travels was China must have also been a central attraction. To the British public, China remained very much the ultimate Other, the antipode to Europe. Travel, to the Victorians, was an exploration of time as well as space. The grand tour and the "Mediterranean passion" ushered British travelers back in time, moving along the corridor of history to Rome, to Athens, to Egypt.[153] Travel literature on China invited the reader to enter a museum of the grotesque, an ancient civilization frozen in time, just as did the "Ten thousand things Chinese" exhibition at the Hyde Park in 1842 (which attracted throngs of visitors, including Thomas and Jane Carlyle), and the Chinese junk brought to London in 1848.[154]

Like narrating, mapping aimed at producing factual knowledge—in this case, often in strictly defined modes of visual representation—of place. As traveling scientists and field naturalists moved from place to place, they connected dots on maps or, in previously unexplored areas, mapped. Mapping turned undefined space into a set of reference points. In the Humboldtian fashion, Victorian explorers, naturalists or otherwise, carried field glasses, barometers, and other survey equipment.[155] They marked down routes, lakes, mountains, and rivers, they measured distances and heights, and they (re)named places. Mapping meant possessing. Landscapes were reduced to signs, lines, and numbers; experiences to abstractions; animals and plants to patches of color in diagrams, thousands of square miles to a piece of paper, portable and transmissible. (Re)naming translated the unfamiliar, the strange, into the familiar. It insisted on disseminating one's existence, one's presence. It struggled to reconfigure the territories, perhaps more symbolic than geopolitical; it asserted, however indirectly, certain power relations. It strove to create an imperial space.[156]

But China was not a void, not a museum, not a blank slate to be written upon at will. Only a century before, the Qing forces swept its western frontiers, established Xinjiang, and thwarted the ambition of the equally aggressive Russian Empire. They brought Jesuit missionaries with them to map the new territories and to depict their victories.[157] And if they did not fully recognize the potential of Western science and technology, if they underestimated the threats from the Western sea powers, it had as much to do with internal political struggle as "veneration of tradition," "ignorance," or any other reasons the Westerners commonly attributed to them.[158]

Our inquiry, smaller and more focused than this broad sketch, told a similarly complex story. In the context of travel and fieldwork, at the level of everyday life, the British in China could not ignore the ever-changing circumstances, the diversity among "the Chinese" they encountered, and the slipperiness of the order they tried to impose on the elements of their fieldwork. Even the discourse of "othering" in travel literature on China admitted ruptures. Not all Chinese were the same. Fortune frequently noticed the marked diversity among the Chinese he came across in different places and constantly shifted his opinions about "the Chinese," despite his many sweeping generalizations. His remarks thus commented on each other.[159]

Travel further into the interior only increased the necessity for naturalists to modify their ideas of China, many of which had been formed from limited experiences in coastal cities, as they witnessed the mosaic of the multiethnic Qing Empire. Similarly, the field naturalists' dependence on the natives' input rendered it difficult to seal the boundary between natural history and folk knowledge. The very action to explore the foreign land and to appropriate the indigenous knowledge entailed the influx of hybrid knowledge that resisted the naturalists' effort to categorize and discipline. Finally, the Chinese informants and collectors did not emerge from the deal as dupes exploited by the naturalists; they had the means to bargain with the naturalists and usually walked off with what they wanted.[160] Without direct evidence, it is hard to tell how they might have viewed their experiences in the field, chasing ugly bugs with a large net or gathering common grasses under a scorching sun. Perhaps like their booted, "red-haired" employers, they also saw it as an adventure. Some, at least.

Epilogue

BY THE END of the nineteenth century, the knowledge of British naturalists about China's natural world had come a long way from that of Joseph Banks and his correspondents in Old Canton. They now knew a great deal about the flora and fauna of China proper and beyond. The days when all they had seen of China was a few streets in Canton seemed a distant memory. British naturalists had collected as far afield as Hainan, Taiwan, Manchuria, Yunnan, and the vast territories in between. They had established extensive networks and regular forums for scientific communication. They had translated and digested much Chinese lore about the natural environment, whether it was folk knowledge or intellectual tradition. In London, British botanists compiled a multivolume work on Chinese plants, for which they received acclaim from their colleagues across continental Europe and the United States. Kew Gardens had amassed an unrivaled collection of Chinese plants, and the British Museum of Natural History had acquired Henry Hance's splendid herbarium. The museum was also well stocked with animal specimens from China. British zoologists now spoke with broad knowledge about Chinese birds and mammals.

Our survey of the naturalists' concerns and methods requires that we remap the intellectual landscape of natural history in the eighteenth and nineteenth century. Historians of natural history have neglected certain currents and research areas that were considered important at the time. Increasing attention to the field tradition in natural history is a welcome development, but the attention has not been extended to folk knowledge. The textual tradition in natural history, well appreciated by students of Renaissance and early modern science, has been all but forgotten by those working on the later period. A reappraisal of these knowledge traditions will restore neglected but pertinent research areas such as the history of cultivated plants to their proper places in history and will help scholars unravel the strands of knowledge in natural history.

This study thus confirms the growing recognition of natural history as an inclusive, complex, extensive, and heterogeneous enterprise. Horticultural interests, for example, propelled much of the botanical investigations of China, and it would not be helpful to treat them as entirely separate enterprises. Private commercial nurseries dealing with exotic flowers, such as James Veitch & Sons, deserve serious scholarly attention, especially when so much ink has been devoted to botanical gardens. A whole class of scientific travelers and professional collectors operated in the large marketplace for specimens and exotic plants. Robert Fortune and Alfred Russel Wallace cofounded a seed company following their careers as professional collectors, and both frequently received requests from the scientific establishment for advice and assistance.[1] In China, a vast community of diplomats, missionaries, and merchants carried out most such research, and their interests and goals did not always conform with those of the scientific establishment in Europe. Natural history cannot be fully understood when severed from this sociocultural continuum.

Natural history aspired to study every object of nature. Its extent was global, and spatiality was an inherent feature of the enterprise. The spatiality of natural history co-evolved with European expansion—this can be seen in the interconnected endeavors of exploration, communication, transportation, and spatial thinking about the distribution of nature's objects. Following this tradition, British naturalists saw China as a *space* to explore and to map. The Chinese, their will, and their society and political institutions, in this view, represented obstacles to the naturalists' quest of complete knowledge of the flora, fauna, and geology of the land called China. But the naturalists also understood that *place* mattered in natural history. Not all places are the same. British naturalists took great interest in places whose topographic, geographic, and climate features promised to offer plants and animals of particular value to Britain and its colonies. Human institutions, too, defined places. China was different from the colonies, and British naturalists had to come to terms with this difference. Their research was very much conditioned by the social and cultural environment of the region of China in which they found themselves.

Recent scholarship has shown how illuminating it can be to write the history of science in the context of European expansion during the eighteenth and nineteenth centuries, variously emphasizing exploration, trade, missionary work, or imperialism. It marks an encouraging step toward incorporating broad historical themes into the history of science. But since this literature has focused exclusively on the internal structures, workings, and imaginings of European imperialism, it remains narrowly Western in its concerns and outlook, albeit with a sense of self-criticism and reflection. This approach may be strengthened if we pay closer attention to the encounters between the Europeans and the indigenous people and their knowledge. The European naturalists themselves recognized the value and contributions of the natives and actively sought their assistance.

The discourse of the naturalists, however, considered the indigenous people merely as helpers or troublemakers whose roles in the narrative were determined by their contributions to or obstructions of the naturalists' goals. That perspective could not have been the natives', and it captures only certain aspects of the relationships in the process. The natives no doubt saw their participation in the activities rather differently, and they acted according to their own interpretation of events. Just as the naturalists pursued scientific research for a variety of reasons, the Chinese entered into the relationship because it promised to deliver reciprocal favors, share mutual hobbies, increase their knowledge, pay well, or add excitement and adventure to life. Whatever the power relations between the Western naturalist and the Chinese might have been, both sides tried to make sense of the encounters. And the relationships between the naturalists and the natives were not necessarily culture- or nation-bound. Historical actors negotiated their identities and roles just as we do.

Throughout this book, I have emphasized the importance of various groups of Chinese in British naturalists' scientific research—merchants, artists, herbalists, hunters, for example. However, I have said relatively little about Chinese scholars of the late Qing. Nor have I said much about the Qing government's attempt to introduce Western science and technology into China. This apparent omission actually reflects a local historical reality. As I mentioned before, British naturalists never met Chinese scholars known for their learning in natural history, though many of them were familiar with and used Chinese works on the subject. It is for a similar reason that I have not discussed at length the introduction of Western natural history into Qing China: Natural history failed to attract the immediate attention of China's reformers and modernizers.

In the wake of the Second Opium War, the Chinese government launched the Self-Strengthening movement and struggled to learn Western military technology and related sciences. Dockyards were built; foundries opened; Western-style schools founded; and translation offices established. John Fryer, William J. P. Martin, Young John Allen, and others worked for the Qing government on translation projects that turned out hundreds of books in science, engineering, and other fields that were considered important to China's reform enterprise. Independently, missionary and other private organizations also sponsored the translation of Western science books. Despite the overall poor quality of the selection and translation of the books, which is only to be expected, as the translators were not experts on the subjects, the books, together with a few periodicals, made available to the Chinese many introductory science texts. At the same time, the Western-style education that appeared in the major cities constituted another channel for introducing Western science. Although some science courses were occasionally offered by missionary schools, it was the new schools sponsored by the Chinese government that provided the bulk of education in science and engineering.

Because the Chinese reform projects focused on engineering and military technology, however, natural history received little emphasis. It was taught at Western-style schools only as part of the basic sciences and only at a rudimentary level. Few of the Westerners employed by the Chinese government as translators and instructors were naturalists or were particularly interested in natural history. Conversely, few noted Western naturalists in China participated in the Qing government's reform efforts. Missionaries translated and taught a little natural history in the tradition of natural theology. The introduction of Western natural history into China seemed to have a slow start. Nevertheless, Chinese intellectuals, including those who had received a traditional education, were beginning to take an interest in Western natural history.[2]

With the collapse of the Qing dynasty in 1911, China disintegrated into political turmoil and war, but from this chaos a new order gradually emerged. The activities of Western naturalists did not suddenly cease with the political upheaval. Exploration, fieldwork, and plant prospecting in the traditional manner carried on into the new era with full momentum. Plant hunters rambled through Yunnan, Sichuan, Tibet, and Gansu. The plants brought back by George Forrest, Earnest H. Wilson, John Reginald Farrer, Frank Meyer, and Francis Kingdon-Ward conquered the hearts of gardeners and agriculturalists across Europe and the United States. Sport and zoological research still went hand in hand. Arthur de Carle Sowerby, heir to Robert Swinhoe's scientific legacy, pursued big-game hunting in Manchuria and Mongolia, collected specimens, and pondered animals as represented in Chinese art and literature. Organized scientific expeditions, hitherto rare, now became more common. Archaeology, geology, and paleontology rose in importance. Europeans were as active as ever; Americans and Japanese rushed ahead. It was a great age of scientific explorations in the interior of China. In the meantime, the China Branch of the Royal Asiatic Society still provided a forum for sinological discussion. Scholars remained fascinated with Chinese literature on plants and animals, and continued their inquiries in the pages of new journals.[3]

Yet, the study of China's natural world was indeed transforming itself. Going, if not yet gone, were the days when everybody, as long as he or she tried hard enough, could contribute something to botany, zoology, and geology. Novelties were no longer just around every corner, but only in distant lands. New discoveries were reserved for experts. Scientific expeditions were staffed with teams of trained specialists. Lone explorers were beginning to look more eccentric than heroic. The science of natural history itself was experiencing profound changes. The field was fragmenting and appeared to many to be aging. New research approaches, particularly laboratory biology, were flourishing at the expense of its influence and appeal, though it was also gaining grounds in other areas, such as ecology. Professionalization of science was squeezing the space for amateurs. To the tough-minded professional scientists, the traditional research goals, methods, and patterns now seemed unfashionable, crude, primitive.[4]

Equally significant was the shift of the institutional base for studying the flora, fauna, and geology of China. The growth of Western-style education and the proliferation of universities and research institutes in Republican China introduced new forces onto the scene. The kinds of investigations were now institutionalized in Chinese educational and research establishments, conducted by foreign and Chinese scientists working there. Many Chinese had earned higher degrees in science in Europe, Japan, and the United States, and they returned to China to teach and carry out research at the universities. The knowledge, data, and modes of research produced by nineteenth-century Western naturalists would become grist for the mill of Chinese scientists and their Western colleagues in the twentieth century.[5] To unfold that history involves a different set of historical considerations, and it merits a volume of its own.

Appendix

Abbreviations

Notes

Index

Appendix: Selected Biographical Notes

The following biographical notes cover the British naturalists in China whose names frequently appear in the text.

Thomas Beale (c. 1775–1842). Merchant. Owner of a fine garden and aviary in Macao. Friend of John Livingstone and John Reeves. In Canton and Macao, 1792–1842.

Cuthbert Collingwood (1826–1908). Fellow of the Linnean Society (FLS). Surgeon and naturalist (particularly interested in botany and marine biology). Cruised on the China Sea and visited Taiwan and South China, 1866–1867.

Alexander Duncan (fl. 1780s). Surgeon to the English East India Company (EIC) in Canton. Collected plants for Joseph Banks. His brother John Duncan (fl. 1790s), who succeeded him in his position in Canton, also collected for Banks.

Charles Ford (1844–1927). FLS. Superintendent, Department of Botanical and Afforestation, Hong Kong, 1871–1902. Close botanical friend of H. F. Hance.

George Forrest (1873–1932). FLS. Professional collector. Collected plants for private nurseries and the Edinburgh Botanic Gardens in southwestern China. 1904–1932. Famous for introducing many rhododendrons into Britain.

Robert Fortune (1812–1880). Plant hunter. Collected plants for the Horticultural Society, EIC, private nurseries, and so on. Traveled in China four times (1843–1845, 1848–1851, 1853–1856, and 1858) and stopped by at China on his way to Japan in 1861. Famous for his books on his travels in China. Introduced the tea plant from China to India.

Henry Fletcher Hance (1827–1886). Ph.D. Giessen. FLS. British Consular Service in China. In China (mostly Whampoa and Amoy), 1844–1886. Most important

British systematic botanist in China. Author of over 200 botanical papers. Owned a herbarium of more than 22,000 species.

William Hancock (1847–1914). Queen's College, Belfast. FLS. Chinese Imperial Maritime Customs Service. Collected plants in China and Taiwan (fl. 1870s–1880s).

Augustine Henry (1857–1930). M.A., Queen's College, Belfast. FLS. Medical officer/commissioner, Chinese Imperial Maritime Customs Service, 1880–1900. Collected plants in central China, Taiwan, and Yunnan.

Alexander Hosie (1853–1925). M.A. Fellow of the Royal Geographic Society. British Consular Service in China, 1876–1909. Collected plants, mainly in Sichuan and Taiwan.

William Kerr (?–1814). Gardener. Collected plants for Kew Gardens in Canton, 1803–1812, before moving to Ceylon as the superintendent of its colonial botanical garden. Generally failed in his duties as a botanical collector in China.

John David Digues La Touche (1861–1935). Ornithologist. Chinese Imperial Maritime Customs Service, 1882–1921. Collected birds mostly in Fujian, Guangdong, and Taiwan.

John Livingstone (?–1829). Surgeon to the EIC. Arrived in China in 1793. Later lived in Macao. Died on the voyage back to China in 1828–1829. Correspondent of the Horticultural Society. Friend of Beale and Reeves.

James Main (c. 1765–1846). Gardener. Collected plants in Canton, 1792–1794, for Gilbert Slater, who was a shareholder of the EIC and owner of a fine garden.

Charles Maries (c. 1851–1902). FLS. Professional plant hunter. Employed by James Veitch and Sons nursery to collect in China and Japan, 1877–1879.

William Frederick Mayers (1839–1878). Eminent sinologist. Joined the British Consular Service in China in 1857 and died in Shanghai in 1878. Collaborated with Hance on a number of botanical and sinological inquiries.

Richard Oldham (1837–1864). Gardener, Kew Gardens. Collected plants for Kew in East Asia, 1861–1862. Broke ties with Kew while in China to pursue a career as independent plant hunter. Traveled with Swinhoe to collect in Taiwan. Died in Amoy, 1864.

Edward Harper Parker (fl. 1860s–1890s). Botanist. British Consular Service, China, 1868–1895. Collected in Guangdong. Close friend and colleague of Hance.

George Macdonald Home Playfair (fl. 1870s–1880s). B.A., Dublin. M.A. British Consular Service, 1872–1910. Botanist. Collected in Hainan.

Antwerp E. Pratt (fl. 1880s–1890s). Professional zoological and entomological collector. Collected insects and reptiles in central and western China, 1887–1890.

John Reeves (1774–1856). Fellow of the Horticultural Society of London (FHS), FLS, Fellow of the Zoological Society of London (FZS). Fellow of the Royal Society (FRS). Tea inspector at the EIC in Canton, 1812–1831. Collected plants for

Joseph Banks and the Horticultural Society. Supervised Chinese artists to draw natural history drawings. Most important British naturalist in China before the Opium War. Father of John Russell Reeves (1804–1877). FRS, FLS, FHS, who also collected for the Horticultural Society, the British Museum, and so on. Merchant. Remained in China, 1827–1857.

Charles Boughey Rickett (fl. 1880s–1900s). Joined the Foochow Branch of the Hong Kong and Shanghai Banking Corporation in 1889 and remained in China until 1904. Collected birds primarily in Fuzhou and the Wuyi Mountains in the province of Fujian.

John Ross (fl. 1870s–1890s). Missionary, United Presbyterian Church of Scotland. Collected plants in Manchuria, 1872.

Theophilus Sampson (1831–1897). Left England and went to sea while a teenager. In various British and Chinese government services at Canton, 1859–1889. Collected plants in Guangdong. Close botanical friend of Hance.

George Thomas Staunton (1781–1859). FRS. FLS. Hon. DCL Oxon. Correspondent of Joseph Banks. Visited China first time with the Macartney Embassy, 1792–1794. With the EIC in Canton, 1799–1817. First British sinologist.

Frederic William Styan (1858–1934). Tea merchant. Zoologist. In China, 1877–1904. Lived mostly in Jiujiang and Shanghai. Collected birds and mammals in the Yangzi basin.

Robert Swinhoe (1836–1877). FZS, FLS, FRS. University of London. British Consular Service in China, 1854–1873. Zoologist. Collected animals and plants in many parts of China proper. Best known for his research on birds and mammals in Taiwan and Hainan.

George Vachell (1799–?). Chaplain, EIC. In Canton, 1828–1836. Collected for John Stevens Henslow, Professor of Botany at Cambridge University.

Ernest Henry Wilson (1876–1930). Professional collector. Gardener, Kew Gardens. Collected plants in southwestern China for Veitch and Sons, 1899–1902, 1903–1905. Later collected in China and Japan for the Arnold Arboretum. Known as "Chinese Wilson."

Abbreviations

BL	British Library
CUL	Cambridge University Library
DTC	Dawson Turner Copies, Joseph Banks Correspondence, Department of Botany, Natural History Museum of London
Kew	Kew Gardens Archives
NBG	National Botanic Gardens (Glasnevin, Ireland)
NHML	Natural History Museum of London
OIOC	Oriental and India Office Collections, British Library
RHS	Royal Horticultural Society of London
RPS	Royal Pharmaceutical Society of London
SOAS	School of Oriental and African Studies, London
Bretschneider, *History*	Emil Bretschneider, *History of European Botanical Discoveries in China*, 2 vols. (Leipzig: Zentral-Antiquariat, 1962 [1898]).
NQ	*Notes and Queries on China and Japan*
JNCB	*Journal of the North China Branch of the Royal Asiatic Society*
Proc. Zool. Soc.	*Proceedings of the Zoological Society of London*
Trans. Hort. Soc.	*Transactions of the Horticultural Society of London*

Notes

Introduction

1. On the Logan riot, see, for example, *The China Mail*, 11, 12, 15 Sept. 1883. The local Chinese were angered by the relatively light sentence resulting from the trial by English law of a British subject who had murdered a Chinese and wounded two others in a dispute. The anger exploded when a Portuguese sailor on a British ship brutally kicked a Chinese into the river, drowning him. The details about the three British botanists are from Ford to William Thiselton-Dyer, 3 October 1883, Kew: Chinese and Japanese Letters, 150 (278–281); Hance to James Britten, 14 July 1884, NHML. BOT. Handwriting Coll.; Knight Biggerstaff, *The Earliest Modern Government Schools in China* (Ithaca: Cornell University Press, 1961), 37; P. D. Coates, *The China Consuls* (Oxford: Oxford University Press, 1988), 202.

2. The Qing dynasty spanned nearly three hundred years, from 1644 to 1911.

3. There have been a few notable books by botanists on their predecessors' endeavors concerning the flora of China. Emil Bretschneider's *History of European Botanical Discoveries in China*, 2 vols. (Leipzig: Zentral-Antiquariat, 1962 [1898]) is monumental and remains indispensable to anyone interested in the subject. E. H. M. Cox, *Plant-Hunting in China: A History of Botanical Exploration in China and the Tibetan Marches* (London: Collins, 1945); Alice M. Coats, *The Plant Hunters* (New York: McGraw-Hill, 1970), published in Britain under the title *The Quest of Flowers*, has a long chapter on Western plant hunters in China. Stephen A. Spongberg, *A Reunion of Trees: The Discovery of Exotic Plants and Their Introduction into North American and European Landscapes* (Cambridge, Mass.: Harvard University Press, 1990) also pays much attention to China. Er-mi Zhao and Kraig Adler, *Herpetology of China* (published by Society for the Study of Amphibians and Reptiles, 1993) includes a historical essay.

4. See, for example, David Elliston Allen, *The Naturalist in Britain: A Social History* (Princeton: Princeton University Press, 1994 [1976]); Lynn Barber, *The*

Heyday of Natural History (New York: Doubleday, 1980); Lynn Merrill, *The Romance of Victorian Natural History* (Oxford: Oxford University Press, 1989); Ann B. Shteir, *Cultivating Women, Cultivating Science: Flora's Daughters and Botany in England, 1760 to 1860* (Baltimore: Johns Hopkins University Press, 1996); N. Jardine, J. A. Secord, and E. C. Spray, eds., *Cultures of Natural History* (New York: Cambridge University Press, 1996), which does not limit its attention to Britain.

5. My definition of the "contact zone" is adapted from Mary Louise Pratt's book *The Imperial Eye: Travel Writing and Transculturation* (London: Routledge, 1992), 6–7. Arif Dirlik has suggested its application in Chinese history in his essay "Chinese History and the Question of Orientalism," *History and Theory* 35 (1996): 96–122, esp. 112–113. See also James Clifford, *Routes: Travel and Translation in the Late Twentieth Century* (Cambridge, Mass.: Harvard University Press, 1997), 188–219. The idea of "borderlands" is widely used in Chicano studies. See, for example, Gloria E. Anzaldua, *Borderlands/La Frontera: The New Mestiza* (San Francisco: Aunt Lute Press, 1987). For a recent discussion on the notion of "borderlands" in American history, see Jeremy Adelman and Stephen Aron, "From Borderlands to Borders: Empires, Nation-States, and the Peoples in Between in North American History." *American Historical Review* 104 (June 1999): 814–841. I have also drawn insights from *Tensions of Empire: Colonial Cultures in a Bourgeois World*, ed. Frederick Cooper and Ann Laura Stoler (Berkeley: University of California Press, 1997); Homi Bhabha, *The Location of Culture* (London: Routledge, 1994); Anne McClintock, *Imperial Leather: Race, Gender and Sexuality in the Colonial Contest* (London: Routledge, 1995).

6. The issue of scientific imperialism is discussed in, for example, Roy MacLeod, "On Visiting the 'Moving Metropolis': Reflections on the Architecture of Imperial Science," in *Scientific Colonialism: A Cross-Cultural Comparison*, ed. Nathan Reingold and Marc Rothenberg (Washington, D.C.: Smithsonian Institution Press, 1987), 217–249; Paolo Palladino and Michael Worboys, "Science and Imperialism," *Isis* 84 (1993): 91–102; Richard Drayton, "Science and the European Empires," *Journal of Imperial and Commonwealth History* 23 (1996): 503–510; William Storey, ed., *Scientific Aspects of European Expansion* (Brookfield: Variorum, 1996), xiii–xxi; and the introduction to Roy MacLeod, ed., *Nature and Empire: Science and the Colonial Enterprise, Osiris* 15 (2000).

7. A few recent works in scientific imperialism emphasize native response and resistance. Deepak Kumar, *Science and the Raj, 1857–1905* (Delhi: Oxford University Press, 1997); Zaheer Baber, *The Science of Empire: Scientific Knowledge, Civilization, and Colonial Rule in India* (Albany: State University of New York Press, 1996); Richard Grove, *Ecology, Climate and Empire: Colonialism and Global Environmental History, 1400–1940* (Cambridge: White Horse Press, 1997). See also David Arnold, *Colonizing the Body: State Medicine and Epidemic Disease in Nineteenth-Century India* (Berkeley: University of California Press, 1993); David Arnold, ed., *Imperial Medicine and Indigenous Societies* (Manchester: Manchester University Press, 1988).

8. I use "natives," "indigenous people," "indigenous knowledge," and other similar terms advisedly, for they seem to imply some kind of cultural purity/simplicity, physical immobility, and even passivity and primitiveness. In a way, they repeat the Western/non-Western dichotomy. For the sake of convenience, however, I will not avoid using them in the book.

9. Paul Cohen, *Discovering History in China: American Historical Writing on the*

Recent Chinese Past (New York: Columbia University Press, 1984). For the new scholarly developments, see, for example, the special issue of *The American Neptune* 48, no. 1 (Winter 1988) on Qing dynasty maritime relations; Aihwa Ong and Donald Nonini, eds., *Ungrounded Empires: The Cultural Politics of Modern Chinese Transnationalism* (New York: Routledge, 1997); Gungwu Wang, *The Chinese Overseas: From Earthbound China to the Quest for Autonomy* (Cambridge, Mass.: Harvard University Press, 2000); James Hevia, *Cherishing Men from Afar: Qing Guest Ritual and the Macartney Embassy of 1793* (Durham: Duke University Press, 1995); the essays in *Modern China* 24, no. 2 (1998); Joanna Waley-Cohen's revisionist survey, *The Sextants of Beijing: Global Currents in Chinese History* (New York: W. W. Norton, 1999); Laura Hostetler, *Qing Colonial Enterprise: Ethnography and Cartography in Early Modern China* (Chicago: University of Chicago Press, 2001), which may be read together with my review of the book in *Metascience* 10 (2001): 458–461. In recent years, historians of China, notably Kenneth Pomeranz, have also revised early modern world history by situating the economic and ecological history of China in global and comparative contexts. Their work is summarized in Robert B. Marks, *The Origins of the Modern World: A Global and Ecological Narrative* (Lanham, Md.: Rowman and Littlefield, 2002).

    10. For discussions on this issue, see, for example, Carlo Ginzburg, *Clues, Myths, and the Historical Method* (Baltimore: Johns Hopkins University Press, 1992) and Ranajit Guha and Gayatri Chakravorty Spivak, eds., *Selected Subaltern Studies* (New York: Oxford University Press, 1988).

1. Natural History in a Chinese Entrepôt

    1. See, for example, Pamela Smith and Paula Findlen, eds., *Merchants and Marvels: Commerce, Science, and Art in Early Modern Europe* (London: Routledge, 2001); Harold Cook, "The Moral Economy of Natural History and Medicine in the Dutch Golden Age," in *Contemporary Explorations in the Culture of the Low Countries*, ed. William Z. Shetter and Inge Van der Cruysse (Lanham, Md.: University Press of America, 1996), 39–47.

    2. On voyages and explorations in the eighteenth century, see David Mackay, *In the Wake of Cook: Exploration, Science and Empire, 1780–1801* (London: Croom Helm, 1985); Marie-Noëlle Bourguet, "Voyage, mer et science au XVIIIe siècle," *Bulletin de la Société d'Histoire Moderne et Contemporaine* 1–2 (1997): 39–56; Jordan Kellman, "Discovery and Enlightenment at Sea: Maritime Exploration and Observation in the Eighteenth-Century French Scientific Community" (Ph.D. diss., Princeton University, 1998); David Philip Miller and Peter Hanns Reill, eds., *Visions of Empire: Voyages, Botany, and Representations of Nature* (Cambridge, England: Cambridge University Press, 1996); Derek Howse, ed., *Background to Discovery: Pacific Exploration from Dampier to Cook* (Berkeley: University of California Press, 1990); Roy MacLeod and Philip Rehbock, eds., *Nature in Its Greatest Extent: Western Science in the Pacific* (Honolulu: University of Hawaii Press, 1988), 13–86; Bernard Smith, *European Vision and the South Pacific*, 2nd ed. (New Haven: Yale University Press, 1985).

    3. Louise E. Robbins, *Elephant Slaves and Pampered Parrots: Exotic Animals in Eighteenth-Century Paris* (Baltimore: Johns Hopkins University Press, 2002), ch. 1; N. A. M. Rodger, *The Wooden World: An Anatomy of the Georgian Navy* (London:

Fantana Press, 1988), 68–71; Wilma George, "Alive or Dead: Zoological Collections in the Seventeenth Century," in *The Origins of Museums: The Cabinet of Curiosities in Sixteenth- and Seventeenth-Century Europe*, ed. Oliver Impey and Arthur MacGregor (Oxford: Clarendon Press, 1985), 179–187; idem, "Sources and Background to Discoveries of New Animals in the Sixteenth and Seventeenth Centuries," *History of Science* 18 (1980): 79–104. On the exchange of plants and animals, see Alfred Crosby, *Ecological Imperialism: The Biological Expansion of Europe, 900–1900* (Cambridge, England: Cambridge University Press, 1986); A. J. R. Russell-Wood, *A World on the Move: The Portuguese in Africa, Asia, and America, 1415–1808* (New York: St. Martin's Press, 1992), 148–182.

4. For example, Paula Findlen, *Possessing Nature: Museums, Collecting, and Scientific Culture in Early Modern Italy* (Berkeley: University of California Press, 1994); Oliver Impey and Arthur MacGregor, eds., *The Origins of Museums*. For the intellectual background, see Lorraine Daston and Katharine Park, *Wonders and the Order of Nature, 1150–1750* (New York: Zone Books, 1998).

5. Steven Shapin, *A Social History of Truth: Credibility and Science in Seventeenth-Century England* (Chicago: University of Chicago Press, 1994), 243–266; Daniel Carey, "Compiling Nature's History: Travellers and Travel Narratives in the Early Royal Society," *Annals of Science* 54 (1997): 269–292; Lisbet Koerner, *Linnaeus: Nature and Nation* (Cambridge, Mass.: Harvard University Press, 1999), which discusses Linnean travel; David Lux and Harold Cook, "Closed Circles or Open Networks?: Communicating at a Distance during the Scientific Revolutions," *History of Science* 36 (1998): 179–211.

6. Richard Grove, *Green Imperialism: Colonial Expansion, Tropical Island Edens and the Origins of Environmentalism, 1600–1800* (Cambridge, England: Cambridge University Press, 1995), ch. 3; Georgius Everhardus Rumphius, *The Ambonese Curiosity Cabinet*, trans. E. M. Beekman (New Haven: Yale University Press, 1999).

7. I have drawn upon a diverse body of literature on cultural encounters. For example, Stuart B. Schwartz, ed., *Implicit Understandings: Observing, Reporting, and Reflecting on the Encounters between Europeans and Other Peoples in the Early Modern Era* (New York: Cambridge University Press, 1994); James Axtell, *After Columbus: Essays in the Ethnohistory of Colonial North America* (Oxford: Oxford University Press, 1988); Nicholas Thomas, *Entangled Objects: Exchange, Material Culture, and Colonialism in the Pacific* (Cambridge, Mass.: Harvard University Press, 1991).

8. For example, Richard Grove, "Indigenous Knowledge and the Significance of South-West India for Portuguese and Dutch Constructions of Tropical Nature," *Nature and the Orient: The Environmental History of South and Southeast Asia*, ed. Richard Grove, Vinita Damodaran, and Satpal Sangwan (Delhi: Oxford University Press, 1998), 187–209.

9. The introduction of plants from Asia to Europe during this period is discussed in Donald F. Lach, *Asia in the Making of Europe*, vol. 2, book 3 (Chicago: University of Chicago Press, 1965), 427–445.

10. The essayist Leigh Hunt (1784–1859) was a witness. He wondered about the Chinese and tea-drinking, and said, "[Their] tea-cup representations of themselves (which are the only ones popularly known), impress us irresistibly with a fancy that they are a people all toddling, little-eyed." Leigh Hunt, *Leigh Hunt Essays (Selected)* (London: Everyman's Library, 1929), 255–259. Ironically, those pictures were of the genre of export painting and were specially designed for the

Western taste. See Chapter 2 for more on Chinese export art. John E. Wills, Jr., provides a broad sketch of Asian-European maritime trade in "European Consumption and Asian Production in the Seventeenth and Eighteenth Centuries," in John Brewer and Roy Porter, eds., *Consumption and the World of Goods* (London: Routledge, 1993), 133–147. See also his "Maritime Asia, 1500–1800: The Interactive Emergence of European Domination," *American Historical Review* 98 (1993): 83–105; Edwin J. van Kley, "Europe's 'Discovery' of China and the Writing of World History," *American Historical Review* 76 (1971): 358–385.

11. The Dutch India Company alone shipped more than three million pieces of chinaware to Europe between 1602 and 1657, not counting those shipped to the colonies and many other parts of the world. Chen Gaohua and Chen Shangsheng, *Zhongguo haiwai jiaotong shi* (Taipei: Wenjin, 1997), 254. The French and the British together shipped more than 1.3 million pieces in 1736 alone. Louis Dermigny, *La Chine et l'Occident, le commerce à Canton au XVIIIe siècle, 1719–1833*, vol. 1 (Paris: S.E.V.P.E.N., 1964), 391. Like Dermigny's work, Hosea Ballou Morse, *The Chronicles of the East India Company Trading to China 1635–1834*, 5 vols. (Oxford: Oxford University Press, 1926–29), hereafter cited as *Chronicles*, also contains much statistical information about the China trade.

12. Donald F. Lach, *Asia in the Making of Europe*, vol. 2, book 1, and vol. 3, books 1 and 4 (vol. 3 was coauthored with Edwin Van Kley); Hugh Honour, *Chinoiserie: The Vision of Cathay* (London: John Murray, 1961); Sprague Allen, *Tides in English Taste (1619–1800): A Background for the Study of Literature*, vol. 1 (New York: Pageant Books, 1958), 180–256; Basil Guy, "The French Image of China before and after Voltaire," *Studies on Voltaire and the Eighteenth Century* 21 (1963); William W. Appleton, *A Cycle of Cathay: The Chinese Vogue in England during the Seventeenth and Eighteenth Centuries* (New York: Columbia University Press, 1951); Patrick Conner, *Oriental Architecture in the West* (London: Thames and Hudson, 1979); Arthur Lovejoy, "The Chinese Origin of a Romanticism," in his *Essays in the History of Ideas* (Baltimore: Johns Hopkins University Press, 1948), 99–135.

13. J. H. Plum, "The Royal Porcelain Craze," in his *In the Light of History* (New York: Delta Publishing, 1972), 57–69; Neil MacKendrick, John Brewer, and J. H. Plumb, *The Birth of a Consumer Society: The Commercialization of Eighteenth-Century England* (Bloomington: Indiana University Press, 1982), 100–145.

14. Osvald Sirén, *China and the Gardens of Europe of the Eighteenth Century* (New York: Ronald Press, 1950); Jurgis Baltrusaitis, "Land of Illusion: China and the 18th Century Garden," *Landscape* 11, no. 2 (1961–62): 5–11. See also the references cited in note 12, most of which also discuss China and European gardens.

15. Francis Hutcheson, *An Inquiry into the Original of Our Ideas of Beauty and Virtue* (1725), in his *Collected Works* (Hildesheim: G. Olms, 1969–71), vol. 3, 76–84; Henry Home Kames, *Elements of Criticism* (New York: Johnson, 1967 [1762]), 316–320.

16. Eileen Harris, "Designs of Chinese Buildings and the Dissertation on Oriental Gardening," in John Harris, ed., *Sir William Chambers* (London: A. Zwemmer, 1970), 144–162. See also R. C. Bald, "Sir William Chambers and the Chinese Garden," *Journal of the History of Ideas* 11 (1950): 287–320. Articles on Chinese gardening and gardens in nineteenth-century British horticultural magazines were simply countless. See, for example, *Gardeners Chronicle*, Loudon's *Gardener's Magazine*, and *The Floricultural Cabinet*. The last mentioned even reprinted

William Chambers's "A Dissertation on Oriental Gardening," which ran in parts from 1838 through 1839.

17. James Main, "Reminiscences of a Voyage to and from China, in the Years, 1792–3–4," *Horticultural Register* 5 (1836): 62–63.

18. "Huadi" was spelled differently in the writings of British travelers and residents in China. Another common rendition was Fa-te.

19. James Cunningham, a surgeon for the East India Company, collected botanical specimens on the Chinese coast for Hans Sloane over several years from 1698 onward, but it was a rather isolated attempt. BL. SL. MS. 4025. ff. 81–89. Franklin P. Metacalf, "Travellers and Explorers in Fukien before 1700," *The Hong Kong Naturalist* 5 (1934): 252–271, esp. 266–268; Bretschneider, *History*, 31.

20. Abigail Lustig, "Cultivating Knowledge in Nineteenth-Century English Gardens," *Science in Context* 13 (2000): 155–181.

21. Harold R. Fletcher, *The Story of the Royal Horticultural Society, 1804–1968* (Oxford: Oxford University Press, 1969), 19–43.

22. Robert Entenmann, "Catholics and Society in Eighteenth-Century Sichuan," *Christianity in China*, ed. Daniel H. Bays (Stanford: Stanford University Press, 1996), 8–23; Eric Widmer, *The Russian Ecclesiastical Mission in Peking during the Eighteenth Century* (Cambridge, Mass.: Harvard University Press, 1976).

23. I use "the Factories" when I refer to the buildings and use "the Canton Factory" for the English East India Company at Canton.

24. For the East India Company in China, see Morse, *Chronicles;* Hoh-cheung Mui and Lorna H. Mui, *The Management of Monopoly: A Study of the East India Company's Conduct of Its Tea Trade, 1784–1833* (Vancouver: University of British Columbia Press, 1984).

25. *The Chinese Repository* 5 (May 1836–1837): 426–432. The numbers included the captains and traders who lived in the Factories during the trade season. There were also dozens of Parsees, who were particularly active in the trade between India and China.

26. On the life of Westerners in Canton, see William C. Hunter, *Bits of Old China* (Taipei: Ch'eng-Wen Publishing Co., 1966 [1855]); idem, *The "Fan Kwae" at Canton before Treaty Days 1825–44* (Taipei: Ch'eng-wen Publishing Co., 1965 [1882]); Jacques Downs, *The Golden Ghetto: The American Commercial Community at Canton and the Shaping of American China Policy, 1784–1844* (Bethlehem, Pa.: Lehigh University Press, 1997), 19–64, which also includes descriptions of the social life of the British traders. S. W. Williams, "Recollections of China prior to 1840," *JNCB*, n.s., 8 (1873): 1–21.

27. On the production and trade of tea, see Robert Gardella, "The Antebellum Canton Tea Trade: Recent Perspectives," *The American Neptune* 48 (1988): 261–270; idem, *Harvesting Mountains: Fujian and the China Tea Trade, 1757–1937* (Berkeley: University of California Press, 1994); Henry Hobhouse, *Seeds of Change: Five Plants that Transformed Mankind* (New York: Harper & Row, 1987), 95–140. On the consumption of tea in Britain, see James Walvin, *Fruits of Empire: Exotic Produce and British Trade, 1660–1800* (New York: New York University Press, 1997); Denys Forrest, *Tea for the British* (London: Chatto and Windus, 1973). William H. Ukers, *All about Tea*, 2 vols. (New York: The Tea and Coffee Trade Journal Company, 1935) is encyclopedic and global in its coverage.

28. For an introductory essay on the Canton trade, see Frederic Wakeman,

"The Canton Trade and the Opium War," in John King Fairbank and Denis Crispin Twitchett, eds., *The Cambridge History of China*, vol. 10, part 1 (Cambridge, England: Cambridge University Press, 1978), ch. 4. Information about the trade items can be found in H. B. Morse, *Chronicles*. On the history of the opium trade, see Su Zhiliang, *Zhongguo dupin shi* (Shanghai: Shanghai renmin chubanshe, 1997), esp. ch. 3. For the East Indiamen and their trade routes, see Jean Sutton, *Lords of the East: The East India Company and Its Ships* (London: Conway Maritime Press, 1981).

29. This is the traditional view. Jacques M. Downs argues that the importance of the fur trade has been overrated. See his *Golden Ghetto*, ch. 3. Whatever the cause might have been, the American merchants ventured into the opium trade in the early nineteenth century.

30. For a discussion on this aspect of Canton, see Fa-ti Fan, "Science in a Chinese Entrepôt: British Naturalists and Their Chinese Associates in Old Canton," *Osiris* 18 (2003): 60-78.

31. For example, John E. Wills, Jr. "Maritime Asia, 1500–1800"; Robert B. Marks, *Tigers, Rice, Silk, & Silt: Environment and Economy in Late Imperial South China* (New York: Cambridge University Press, 1998), 163–194; Jennifer Wayne Cushman, *Fields from the Sea: Chinese Junk Trade with Siam during the Late Eighteenth and Early Nineteenth Century* (Ithaca: Cornell University Press, 1993); Dian Murray, *Pirates of the South China Coast, 1790–1810* (Stanford: Stanford University Press, 1987).

32. Huang Qicheng and Huang Guoxing, *Guangdong shangbang* (Hong Kong: Zhonghua shuju, 1995).

33. John Robert Morrison, *A Chinese Commercial Guide, Consisting of a Collection of Details Respecting Foreign Trade in China* (Canton, 1834) explains the official procedures of checking in with the Chinese authorities when the ships arrived. Personal observations can be found in Peter Quennell, ed., *The Prodigal Rake: Memoirs of William Hickey* (New York: E. P. Dutton, 1962), 132–137; Hunter, *The "Fan Kwae" at Canton*, 11–14, 20–25; C. Toogood Downing, *The Fan-Qui in China, in 1836–7*, vol. 1 (London: Henry Colburn Publisher, 1838). See also H. B. Morse, *The International Relations of the Chinese Empire*, vol. 1, ch. 4 (Taipei: Wenxing, 1963 [1910–18]).

34. Nan Powell Hodges, *The Voyage of the* Peacock: *A Journal by Benajah Ticknor, Naval Surgeon* (Ann Arbor: University of Michigan Press, 1991), 171.

35. Downing, *The Fan-Qui in China*, vol. 2, 68.

36. Ibid., 66.

37. Many contemporary witnesses described that district of Canton in detail. I have drawn upon, among others, George Bennett, *Wanderings in New South Wales, Batavia, Pedir Coast, Singapore, and China; Being the Journal of a Naturalist in Those Countries, during 1832, 1833, and 1834*, vol. 2 (London: Richard Bentley, 1834), 86–116; Hunter, *Bits of Old China*, 4–7, 12–15; idem, *The "Fan-Qae" at Canton*, 20–26; Downing, *The Fan-Qui in Chin*, vol. 2, 40–57, 211–214, 230–232. H. A. Crosby Forbes, *Shopping in China: The Artisan Community at Canton, 1825–1830* ([Washington, D.C.]: International Exhibitions Foundation, 1979) is an exhibition catalog of drawings of the shops of different trades in Old Canton.

38. *The Canton Register*, no. 17, April 26, 1828.

39. There were a few notable incidents like the *Lady Hughes* and the *Topaze* controversy, but they were resolved swiftly by both the Chinese and the British.

Details of the two events can be found in Morse, *Chronicles*, vol. 2, 99–109; vol. 4, 18, 27–41.

40. Bretschneider, *History*, vol. 1 covers most of the Jesuit missionaries who contributed to the research into China's natural history in the seventeenth and eighteenth centuries. Marie-Pierre Dumoulin-Genest, "Note sur les plantes chinoises dans les jardins français du XVIIIe siècle: De l'expérimentation à la diffusion," *Études chinoises* 11, no. 2 (1992): 141–158; idem, "L'introduction et l'acclimatation des plantes chinoises en France au XVIIIe siècle" (Ph.D. diss., Écoles des Hautes Études en Sciences Sociales, 1994).

41. J. L. Cranmer-Byng, "The First English Sinologists: Sir George Stanton and the Reverend Robert Morrison," in *Symposium on Historical, Archaeological and Linguistic Studies on South China, South-East Asia and the Hong Kong Region*, ed. F. S. Drake and Wolfram Eberhard (Hong Kong: Hong Kong University Press, 1967), 247–259; Susan Reed Stifler, "The Language Students of the East India Company's Canton Factory," *JNCB* 69 (1938): 46–82; T. H. Bartlett, *Singular Listlessness: A Short History of Chinese Books and British Scholars* (London: Wellsweep, 1989), 19–58.

42. Two recent major works on the Macartney Embassy are Alain Peyrefitte, *The Immobile Empire* (New York: Knopf, 1992), which is a detailed narrative, and James Hevia, *Cherishing Men from Afar: Qing Guest Ritual and the Macartney Embassy of 1793* (Durham: Duke University Press, 1995), which is a "postmodern" study. Both have stirred some controversy. Compare Harriet T. Zurndorfer, "La sinologie immobile," *Études chinoises* 8 (1989): 99–120 and Alain Peyrefitte's reply in the same journal, vol. 9, 242–248. On Heavia's book: Pamela Kyle Crossley's review in *Harvard Journal of Asiatic Studies* 57 (1997): 597–611. See also the exchange between Joseph Esherick and Hevia in *Modern China* 24 (1997): 135–161, 319–327.

43. The naturalists on the two missions, George L. Staunton and Clarke Abel, respectively, wrote extensively about Chinese husbandry, gardening, and natural history. See G. L. Staunton, *An Authentic Account of Embassy from the King of Great Britain to the Emperor of China*, 2 vols. (London: Stockdale, 1797); Clarke Abel, *Narrative of a Journey in the Interior of China in the Years 1816–1817* (London: Longman and Hurst, 1818). For their natural historical research, see Bretschneider, *History*, 156–163, 225–237. The scientific aspect of the Macartney mission is noticed in J. L. Cranmer-Byng and Trevor H. Levere, "A Case Study in Cultural Collision: Scientific Apparatus in the Macartney Embassy to China, 1793," *Annals of Science* 38 (1981): 503–525, which discusses the scientific instruments brought with the mission. For a different perspective, see Joanna Waley-Cohen, "China and Western Technology in the Late Eighteenth Century," *American Historical Review* 98, no. 5:1525–1544.

44. There was at least one group of Britons who traversed one part of China about this time. They were some shipwreck victims who had to walk across half the province of Guangdong to Canton and who were at one point forced by circumstances to exhibit themselves to Chinese spectators for money. J. R. Supercargo, *Diary of a Journey Overland, through the Maritime Provinces of China, from Manchao, on the South Coast of Hainan, to Canton, in the Years 1819 and 1820* (London: Printed for Sir Richard Phillips and Co., 1822).

45. The quote is from f. 68 of Joseph Banks's instructions for William Kerr in DTC, vol. 14, ff. 61–68. See also Joseph Banks, "Hints on the Subject of Gar-

dening, suggested to the Gentlemen who attend the Embassy to China," The Linnean Society. MSS. Banks's instructions for Clarke Abel are in DTC, vol. 19, ff. 239–245; those for James Hooper, the gardener-naturalist on the mission, are collected in DTC, vol. 19, ff. 225, 231–232.

46. DTC, vol. 19, f. 240.

47. Ibid., ff. 239–245. The quote is from f. 242.

48. Joseph Sabine, "Account of Several New Chinese and Indian Chrysanthemums . . . ," *Trans. Hort. Soc.*, sr. 1, 6 (1826): 326.

49. Bretschneider, *History*, 211–215.

50. On the popularity of natural history in eighteenth- and nineteenth-century Britain: Lynn Barber, *The Heyday of Natural History* (New York: Doubleday, 1980); David Elliston Allen, *The Naturalist in Britain: A Social History* (Princeton: Princeton University Press, 1994 [1976]); Lynn L. Merrill, *The Romance of Victorian Natural History* (New York: Oxford University Press, 1989); Ann B. Shteir, *Cultivating Women, Cultivating Science: Flora's Daughters and Botany in England, 1760 to 1860* (Baltimore: Johns Hopkins University Press, 1996); Barbara T. Gates, *Kindred Nature: Victorian and Edwardian Women Embrace the Living World* (Chicago: University of Chicago Press, 1998); N. Jardine, J. A. Secord, and Emma Spary, eds., *Cultures of Natural History* (New York: Cambridge University Press, 1996). Keith Thomas's classic *Man and the Natural World: Changing Attitudes in England, 1500–1800* (London: Allen Lane, 1983) provides a broad cultural background. On British naturalists in other parts of the world, see, for example, Clare Lloyd, *The Traveling Naturalists* (Seattle: University of Washington Press, 1985); Colin Finney, *Paradise Revealed: Natural History in Nineteenth-Century Australia* (Melbourne: Museum of Victoria, 1993); Peter Raby, *Bright Paradise: Victorian Scientific Travellers* (Princeton: Princeton University Press, 1996); Satpal Sangwan, "Natural History in Colonial Context: Profit or Pursuit? British Botanical Enterprise in India, 1778–1820," in *Science and Empires: Historical Studies about Scientific Development and European Expansion*, ed. Patrick Petitjean, Cathrine Jami, and Anne Marie Moulin (Dordrecht: Kluwer Academic Publishers, 1992), 281–298; Deepak Kumar, "The Evolution of Colonial Science in India: Natural History and the East India Company," in John MacKenzie, ed., *Imperialism and the Natural World* (Manchester: Manchester University Press, 1990), 51–67.

51. G. Métailié, "Sir Joseph Banks—an Asian Policy," in R. E. R. Banks et al., eds., *Sir Joseph Banks: A Global Perspective* (Kew: Royal Botanic Gardens, 1994), 157–170; Miller and Reill, eds., *Visions of Empire*, especially the essays in part I; Harold B. Carter, *Sir Joseph Banks, 1743–1820* (London: British Museum, 1988), 290–291, 407, 440–441; John Gascoigne, *Joseph Banks and the English Enlightenment: Useful Knowledge and Polite Culture* (New York: Cambridge University Press, 1994), 179–181, 214; idem, *Science in the Service of Empire: Joseph Banks, the British State and the Uses of Science in the Age of Revolution* (New York: Cambridge University Press, 1998), 140–142.

52. BL. BM. Add. MS. 33978. 177 and Duncan to Banks, 7 March 1789, in the same folder.

53. BL. BM. Add. MS. 33978. 276–279; MS. 33979, 69–70.

54. Staunton, *Memoirs*, 2, 12.

55. George Thomas Staunton, *Memoirs of the Chief Incidents of the Public Life of Sir George Thomas Staunton* (London: Printed for private circulation, 1856), 25.

56. Ibid., 7.
57. OIOC. MSS. EUR. D562/16.
58. Bretschneider, *History*, 225, 256–263.
59. Staunton, *Memoirs*, 54.
60. DTC, vol. 18, f. 195. Reeves might not be as ignorant of botany as he made himself out to be—he was writing to Joseph Banks, the dean of British botany, after all—but then, there is no evidence to prove that his botany was more than rudimentary at that time.
61. Millet to William Hooker, 11 October 1832, Kew: Directors Correspondence 53 (86).
62. Vachell to Henslow, 26 January 1830, CUL. Letters to John Stevens Henslow. Add. 8176. No. 185. Vachell also presented the Cambridge Philosophical Society "about a hundred Chinese fishes." P. J. P. Whitehead and K. A. Joysey, "The Vachell Collection of Chinese Fishes in Cambridge," *The Bulletin of the British Museum (Natural History), Zoological Series* 15, no. 3 (1967): 123–160.
63. Robert Morrison, *Memoirs of the Life and Labours of Robert Morrison*, comp. Eliza Morrison, vol. 2 (London, 1839), 76–77.
64. For example, John Livingstone, "Account of the Method of Dwarfing Trees and Shrubs, as Practised by the Chinese, including their Plan of Propagating from Branches," *Trans. Hort. Soc.* 4 (1822): 224–231; idem, "Account of a Method of ripening Seeds in a wet Season; with some Notices of the Cultivation of certain Vegetables and Plants in China," *Trans. Hort. Soc.* 3 (1822): 183–186; idem, "On the State of Chinese Horticulture and Agriculture; with an Account of several Esculent Vegetables used in China," *Trans. Hort. Soc.* 5 (1824): 49–55.
65. *The Chinese Repository* 4 (May 1835–April 1836): 96–97; Dorothea Scott, "The Morrison Library, an Early Nineteenth Century Collection in the Library of the University of Hong Kong," *Journal of the Hong Kong Branch of the Royal Asiatic Society* 1 (1960–61): 50–67.
66. Morrison, *Memoirs*, vol. 2, 20–21, 29, 76; Reeves to Joseph Banks, 24 March 1820, BL. BM. Add. MS. 33982. 213–216.
67. The quotes are from *The Canton Register*, vol. 2, no. 5, 2 March 1829. On the museum, see also Morrison, *Memoirs*, vol. 2, 427.
68. For example, Ronald Meek, *Social Science and the Ignoble Savage* (Cambridge, England: Cambridge University Press, 1976); Istvan Hont and Michael Ignatieff, eds., *Wealth and Virtue: The Shaping of Political Economy in the Scottish Enlightenment* (Cambridge, England: Cambridge University Press, 1983).
69. For more on Livingstone, Beale, and Reeves, see Chapter 2.
70. Ray Desmond, "Strange and Curious Plants (1772–1820)," in *Plant Hunting for Kew*, ed. F. Nigel Hepper (London: HMSO, 1989), 1–10; idem, *Kew: The History of the Royal Botanical Gardens* (London: Harvell, 1995), 113–126; Fletcher, *The Story of the Royal Horticultural Society*, 91–111.
71. Joseph Poole, who came to Canton to collect plants for the nursery firm Barr and Brookes in 1818, was the only botanical collector who had previously been to China. He had accompanied Clarke Abel, his brother-in-law, on the Amherst Embassy. Bretschneider, *History*, 221; *Trans. Hort. Soc.* 4 (1822): 333; Abel, *Narrative of a Journey in the Interior of China*, vi.
72. DTC, vol. 19, ff. 56–63, on f. 60.
73. RHS. MSS. John D. Parks, "Unpublished Journal, 1823," 1–2. On the importance of carpenters on ship: Rodger, *The Wooden World*, 23.

74. Thomas Manning to Banks, 16 April 1807. BL. BM. Add. MS. 33981. 248–249. Manning, a good friend of the essayist Charles Lamb, was sympathetic toward Kerr, but Manning himself was an eccentric guest who very much annoyed his hosts.

75. DTC, vol. 14, f. 62. Kerr was put "wholly under the protection & intirely under the command of David Lance, Esq, chief Supercargo of the English Factory in China." Banks also requested G. T. Staunton to offer assistance to Kerr, but for some reason, Staunton told Banks that "I can at present, I fear, aid [him] only in inclination." DTC, vol. 16, f. 232.

76. John Livingstone, "John Livingstone's Letter to the Horticultural Society of London," *The Indo-Chinese Gleaner*, vol. 2, no. 9 (1819): 126–131, esp. 127–128. An abridged version of the letter appeared in *Trans. Hort. Soc.* 3 (1822): 421–429.

77. RHS. MSS. John Damper Parks, "Unpublished Journal, 1823," Letter, 20 November 1823.

78. RHS. MSS. John Potts, "Unpublished Journal, 1821–22." See the copy "Fair Journal," under the entry "Observations on Chinese Gardening." Potts said that Beale sent "his gardener up the Country to collect plants where the English are not permitted to go."

79. Main, "Reminiscences of a Voyage to and from China," 172–173.

80. Parks, "Unpublished Journal, 1823," 120–124. Similarly, John Potts observed the techniques of the gardeners at Fa-tee and concluded that they were "rough and awkward." "Indeed," he asserted, "there is no neatness to be observed in the Chinese Gardener, which answers at once for their failure of encreasing [sic] anything delicate or tender." Still, he carefully put down in his journal many Chinese methods of gardening. Potts, "Unpublished Journal, 1821–22 (rough journal)," entries under 22 November 1821, "Method of Propagating," and many other places.

81. The Chinese collectors gathered plants in the hills for them. See, for example, Potts, "Unpublished Journal, 1821–22," entries on Nov. 24, Dec. 7, 19.

82. Findlen, *Possessing Nature*, 170–179; idem, "Inventing Nature: Commerce, Art, and Science in the Early Modern Cabinet of Curiosities," in Smith and Findlen, eds., *Merchants and Marvels*, 297–323.

83. This mode of research remained important to British naturalists in China even in the second half of the nineteenth century. Robert Swinhoe, a well-known zoologist in China, obtained many specimens of fish, frogs, and so on in the markets in Shanghai and other places. "I have been almost daily to the market," he wrote in a letter, "and have bought up nearly everything that was offered for sale." Swinhoe to A. Günter, 12 April 1873, NHML, Z. Keeper's Arch. 1.8. Letters. 1858–75. SM-Z, No. 287. When Arthur Adams, a marine zoologist, was in Hong Kong in the 1840s, he frequented the fishmongers' shops. Arthur Adams, *Travels of a Naturalist in Japan and Manchuria* (London: Hurst and Blackett, 1870), 60–63. The authors of a manual for traveling naturalists emphasized that "[t]he markets must be frequented and searched." Arthur Adams, William B. Baikie, and Charles Barron, *A Manual of Natural History for the Use of Travellers* (London: John Van Voorst, 1854), 656–662.

84. Peter Osbeck, Carl Linnaeus's student to China, described the hundreds of plants, animals, and curious objects he encountered in the marketplaces during his stay in Canton in 1751. Peter Osbeck, *A Voyage to China and the East Indies*, trans. John Reinhold Forster, vol. 1 (London: Benjamin White, 1771), 319–376.

85. There is a fine description of Chinese bird shops in Hong Kong in

Cuthbert Collingwood, *Ramble of a Naturalist on the Shores and Waters of the China Sea* (London: John Murray, 1868), 319–321. Cantonese artists for export painting often depicted street scenes, and the bird-seller was one of their subjects. See the illustrations in Carl Crossman, *The China Trade: Export Paintings, Furniture, Silver and Other Objects* (Princeton, N.J.: Pyne Press, 1972), 102; Downs, *The Golden Ghetto*, 34.

86. On the animals: Downing, *The Fan-Qui in China*, vol. 1, 312–314.

87. Ibid., vol. 2, 234.

88. Robbins, *Elephant Slaves and Pampered Parrots*, ch. 1. When Arthur Adams visited Japan in 1859, he bought from the natives many animals, including a couple of giant salamanders and two bears, which were subsequently donated to the Zoological Society of London. Arthur Adams, *Travels of a Naturalist in Japan and Manchuria*, 277–279, 305–308.

89. *Voyage à Canton, Capitale de la Province de ce nom, à la Chine* . . . , by C. Charpentier Cossigny (Paris: André, 1800), 131–132. Not all animals found at the Canton markets were native to China. The "ourang outang" for sale, which was not necessarily the same kind of animal now bearing the name, might have been brought over from Southeast Asia. According to Peter Osbeck, parrots were among the major items in the Chinese trade with other parts of Asia. Osbeck, *A Voyage to China and the East Indies*, vol. 1, 257. Local histories of Canton all have lists of exotic animals. For example, Qu Dajun, *Guangdong xinyu* (Taipei: Taiwan xuesheng shuju, 1968 [1700]), 1087–1095, 1143, and numerous other places. See also *Guangdong tongzhi* (Taipei, 1968 [1822]), 1720–1721. It may be worth mentioning that the Chinese did not always appreciate Western pets. One of the comments cited in many local histories was on foreign dogs. It described one of the Western dogs as "vicious-looking." The Chinese author, moreover, was puzzled that Westerners (or barbarians) in Macao valued highly a "short and small" dog, "shaped like a lion," because the dog had no particular skills. He also commented that the Westerners valued their dogs more than their "black slaves" (*heinu*). The Westerners ate and slept with their dogs and gave them the best food. He added that there was thus a saying to the effect that one would be better off as a dog than a slave of the barbarians. See Qu, *Guangdong tongzhi*, 1721.

90. *Guangdong xinyu, juan* 24, "*dahudie.*"

91. Downing, *The Fan-Qui in China*, vol. 2, 50–51.

92. J. O. Westwood, "Preface" to E. Donovan, *Natural History of the Insects of China*, new ed. (London: Henry G. Bohn, 1842). The insect boxes contained also dried small fish and probably shells as well. John Richardson, a distinguished ichthyologist, found in the boxes many interesting fish. John Richardson, "Report on the Ichthyology of the Seas of China and Japan," *Report of the 15th Meeting of the British Association for the Advancement of Science* (1845): 187–320. See, for example, pp. 202–203, 213–217.

93. *The Canton Register*, 2 March 1829. H. E. Strickland, a major ornithologist in Britain, observed, "It seems strange that so little has yet been done to obtain specimens of Chinese zoology through the medium of the natives. Thousands of bird-skins are annually sent to Europe by the natives of Brazil, Senegal and Malacca, and there can be no reason why a similar trade should not be established with China. All that the Chinese want is a little instruction in the art of preserving specimens, which might be easily communicated if some of the merchants con-

nected with the tea trade would take an interest in the subject." *Annals and Magazine of Natural History* 12 (1843): 222.

94. More on Beale and his aviary in Chapter 2.

95. George Bennett, *Wanderings in New South Wales*, vol. 2, 57, 58, 61, 68; George Vachell to Henslow, 30 January 1831, CUL. Letters to John Stevens Henslow. Add. 8176. No. 190. Westerners at that time termed the vast space between China proper and Siberia "Tartary."

96. David Low, *On the Domesticated Animals of the British Islands: Comprehending the Natural and Economical History of Species and Varieties; the Description of the Properties of External Form; and Observations on the Principles and Practice of Breeding* (London: Longman, Brown, Green, and Longmans, 1846), 426.

97. Records of the donations and purchases of the animals are in the Zoological Society of London. MSS. Daily Occurrences. For example, in 1831, the Society received a Chinese vulture, two white Chinese geese, four water tortoises, and several pheasants, among others. The Zoological Society of London urged British residents in China to export to Britain "Monkeys, Deer, and all wild Quadrupeds" and "Pheasants of all kinds, except Gold and Silver, Mandarin and other Teal, Fishing Pelicans." Evidently they knew little about Chinese animals, for their lists of animals to be introduced from the colonies were relatively specific and extensive. See *Report of the Committee of Science and Correspondence to the Council of the Zoological Society, March 22nd., 1831* (London, 1831), 11. For the attempts of the Zoological Society of London to collect exotic animals, see R. Fish and I. Montagu, "The Zoological Society and the British Overseas," *Symp. Zool. Soc. Lond.*, no. 40 (1976): 17–48; Adrian Desmond, "The Making of Institutional Zoology in London, 1822–1836," *History of Science* 23 (1985): 153–185, 223–250; Harriet Ritvo, *The Animal Estate: The English and Other Creatures in the Victorian Age* (Cambridge, Mass.: Harvard University Press, 1987), ch. 5. See also Christine Brandon-Jones, "Edward Blyth, Charles Darwin, and the Animal Trade in Nineteenth-Century India and Britain," *Journal of the History of Biology* 30 (1997): 145–178. Robert Jones, "'The Sight of Creatures Strange to Our Climate': London Zoo and the Consumption of the Exotic," *Journal of Victorian Culture* 2, no. 1 (1997): 1–26. The activity was not limited to Britain. For example, R. J. Hoage and William A. Deiss, *New Worlds, New Animals: From Menagerie to Zoological Park in the Nineteenth Century* (Baltimore: Johns Hopkins University Press, 1996); Michael Osborne, *Nature, the Exotic, and the Science of French Colonialism* (Bloomington: Indiana University Press, 1994).

98. On Millet, see Bretschneider, *History*, 298–301. The donations of John and J. R. Reeves to the Museum are recorded in *The History of the Collections Contained in the Natural History Departments of the British Museum* (London: The Trustees of the British Museum, 1904–12), vol. 2, in the sections on mammals and birds.

99. Zhang Dai, *Taoan mengyi* (Taipei: Jingfeng chubanshe, 1989 [1646?]), 88–89. The flower was rare in the south to the Yangzi River.

100. Qiu Chishi (d. 1800), *Yangcheng guchao* (Taipei: Wenhai chubanshe, 1969), 704–706; Qu Dajun, *Guangdong xinyu*, 117; Huang Foyi, *Guangzhou chengfang zhi*, reprint (Guangzhou, 1948), *juan* 5:5, 38.

101. *The Indo-Chinese Gleaner* 2 (July 1819), 127.

102. Shen Fu, *Fusheng liuji* (c. 1810). I use for citation the translation made by Shirley M. Black, *Six Chapters from a Floating Life: The Autobiography of a Chinese Artist* (Oxford: Oxford University Press, 1960), 47.

103. [James] Main, "Observations on Chinese Scenery, Plants, and Gardening, made on a Visit to the City of Canton and its Environs, in the Year 1793 and 1794 . . . ," *Gardener's Magazine* 2 (1827): 139. A hundred Spanish dollars was slightly less than £20. See also Jack Goody, *The Culture of Flowers* (Cambridge, England: Cambridge University Press, 1993), 387–414.

104. Alexander Duncan to Banks, 29 December 1791, BL. BM. Add. MS. 33979. 121–122; Main, "Reminiscences of a Voyage to and from China," 176.

105. Public Record Office, Kew: FO 1048/27/13.

106. Hunter, *Bits of Old China*, 7–12.

107. Main, "Reminiscences of a Voyage to and from China," 148; Hunter, *Bits of Old China*, 8.

108. Bennett, *Wanderings in New South Wales*, 89.

109. Main, "Reminiscences of a Voyage to and from China," 148–149.

110. Reeves's remarks are quoted from *Gardener's Magazine* 11 (1835): 112.

111. Goody, *The Culture of Flowers*, 347–386.

112. Max Weber, *The Religion of China: Confucianism and Taoism* (New York: Free Press, 1968), 234–235.

113. Liang Jiabin, *Guangdong shisan hang kao* (Shanghai: Commercial Press, 1936); Ann Bolbach White, "The Hong Merchants of Canton" (Ph.D. diss., University of Pennsylvania, 1967); Ch'en Kuo-tung, *The Insolvency of the Chinese Hong Merchants, 1760–1834* (Nankang, Taiwan: Institute of Economics, Academic Sinica, 1990); Weng Eang Cheong, *The Hong Merchants of Canton: Chinese Merchants in Sino-Western Trade* (Richmond, England: Curzon Press, 1997), which is comprehensive on the eighteenth century.

114. Weng Eang Cheong, *The Hong Merchants of Canton*, 159–164.

115. Quennell, ed., *The Prodigal Rake*, 143.

116. The French and British generally liked him. Some American traders did not like him, but then Consequa was brought down by the unpaid debts of an American trader. Frederic D. Grant, Jr., "The Failure of the Li-ch'uan Hong: Litigation as a Hazard of Nineteenth-Century Foreign Trade," *The American Neptune* 48 (1988): 243–260.

117. OIOC. IOR. Neg. 11666. "Memoir of the Life of Mr Charles Molony," 11.

118. Harriet Low Hillard, *My Mother's Journal*, ed. Katherine Hillard (Boston: G. H. Ellis, 1900), 38.

119. Morrison, *Memoirs*, vol. 1, 468.

120. OIOC. IOR. Neg. 11666, 11; Hunter, *Bits of Old China*, 43; Morrison, *Memoirs*, 468; White, "The Hong Merchants of Canton," 95–100.

121. Hunter, *Bits of Old China*, 45–46.

122. Hunter, *The Fan Kwae at Canton*, 48; idem, *Bits of Old China*, 80.

123. Descriptions and drawings of the Hong merchants' gardens can be found in Hunter, *Bits of Old China*, 78–79; Josiah Quincy, *The Journals of Major Samuel Shaw* (Boston: W. Crosby and H. P. Nichols, 1847), 179; Main, "Observations on Chinese Scenery, Plants, and Gardening," 136–138; *Tingqua's China*, exhibition 3–22 March 1986 (London: Martyn Gregory, 1986), nos. 95–98.

124. For example, Craig Clunas, *Fruitful Sites: Garden Culture in Ming Dynasty China* (London: Reaktion Books, 1996); Wang Yi, *Yuanlin yu Zhongguo wenhua* (Shanghai: Shanghai renmin chubanshe, 1990); Paolo Santangelo, "Ecologism versus Moralism: Conceptions of Nature in Some Literary Texts of Ming-Qing

Times," in Mark Elvin and Liu Ts'ui-jung, eds., *Sediments of Time: Environment and Society in Chinese History* (Cambridge, England: Cambridge University Press, 1998), 617–657; Joanna F. Handlin Smith, "Gardens in Ch'i Piao-chia's Social World: Wealth and Values in Late-Ming Kiangnan," *Journal of Asian Studies* 51 (1992): 55–81. Norah Titley and Frances Wood, *Oriental Gardens* (London: The British Library, 1991), 77–99, and Maggie Keswick, *The Chinese Garden* (London: Academic Editions, 1978) are beautifully illustrated.

125. The rich enjoyed extravagances, and the poor but cultured made do with whatever was available. The ill-fated Shen Fu tells us in his memoirs, "Simple though my life has been, I have always had a vase of flowers on my table." Because he was too poor to afford a garden, he and his wife built a miniature one in a large pot and imagined themselves living in a fairyland. Shen, *Six Chapters from a Floating Life*, 61–62.

126. Huang Foyi, *Guangzhou chengfang zhi, juan* 5:53–56.

127. Ch'en Kuo-tung, "Pan Youdu (Panqiguan er shi): yiwei chenggong de yanghang shangren," in *Zhongguo haiyang fazhanshi lunwen ji* (Taipei: Academia Sinica, 1993), vol. 5, 245–300. Arthur Hummels, ed., *Eminent Chinese of the Ch'ing Period*, vol. 2 (Washington, D.C.: Library of Congress, 1944), 605, 877.

128. Hunter, *Bits of Old China*, 78.

129. *Guangdong xinyu, juan* 25, *"mudan."*

130. BL. BM. Add. MS. 33979. 84–85; Add. MS. 33978. 113; Add. MS. 33978. 115.

131. Main, "Observations on Chinese Scenery," 135–140.

132. BL. BM. Add. MS. 33981. 227–228; Add. MS. 33981. 231–232.

133. BL. BM. Add. 33981. 227–228; Add. MS. 33981. 229–30; DTC, vol. 17, f. 35. The dwarf tree was eventually presented to the Queen for "Her Majesty's inspection of the Art peculiar to the Chinese, of dwarfing into the picturesque the most lofty Tree of the Forest." BM. Add. MS. 33981. 261.

134. DTC, vol. 18, ff. 193–194.

135. John Potts, "Unpublished Journals, 1821–22," the entries on 12, 13, 14 November 1821, and many others.

136. There are as yet no systematic studies of intercultural gift exchange in the social world of the China trade. Anthropologists have paid certain attention to the exchange of gifts and favors in contemporary Chinese society. Their works offer some insights into the traditional patterns of exchange, especially the connection between gift relations and social relations (in other words, *guanxi*), in Chinese society. See, for example, Mayfair Mei-hui Yang, *Gifts, Favors, and Banquets: The Art of Social Relationships in China* (Ithaca: Cornell University Press, 1994); Andrew Kipnis, *Producing Guanxi: Sentiments, Self, and Subculture in a North China Village* (Durham: Duke University Press, 1997); Yunxiang Yan, *The Flow of Gifts: Reciprocity and Social Networks in a Chinese Village* (Stanford: Stanford University Press, 1996).

137. Hunter, *Bits of Old China*, 31.

138. *Gardener's Magazine* 2 (1827), 422; 11 (1835): 111–112.

139. Ibid., 11 (1835): 437–438.

140. DTC, vol. 19, ff. 123–124, 245.

141. The plant hunter Robert Fortune, who went to China in 1843, presented a Chinese nurseryman in Canton at least a plant he took out with him. It was likely that he was simply following a common practice. Robert Fortune, *Three Years'*

Wanderings in the Northern Provinces of China (London: John Murray, 1847), 95. The Chinese usually planted their flowers in pots, and this mode of gardening probably determined to some extent their preference for flowers. A British observer commented that "the Chinese do not value those flowers which group well, forming massy patches, but only those which show well in a pot." He continued, "They have found fault with several of our European introductions . . . because of their tendency to form large and unwieldy plants." *Gardeners Chronicles,* January 7, 1860, 7.

142. *Gardener's Magazine* 11 (1835): 112.

143. BL. BM. Add. MS. 33978. 276–279.

144. BL. BM. Add. 33979. 121–122.

145. BL. BM. Add. MS. 33978. 112; Add. MS. 33978. 165; Add. MS. 33978. 177.

146. BL. BM. Add. MS. 33978. 211–212.

147. Morrison, *Memoirs,* vol. 2, 424. The annual subscription was thirty Spanish dollars or less than £10 per year.

148. Sutton, *Lords of the East,* ch. 8.

149. BM. Add. MS. 33978. 276–279.

150. RHS. MSS. John Damper Parks, "Unpublished Journal, 1823," 5–9.

151. Alice Coats, *The Plant Hunters,* 94–95; W. Kerr, "Botanical Mission to the Island of Luconia, in the Year 1905," NHML. Botany Library. B: MSS. Kerr. The loss of his Manila plants occurred at Macao in September. Bretschneider, *History,* 270.

152. Joseph Banks, "Instructions for James Smith & George Austin, the Two Gardeners," in DTC, vol. 6, ff. 196–202; John Livingstone, "Observations on the Difficulties which Have Existed in the Transportation of Plants from China to England, and Suggestions for Obviating Them," *Trans. Hort. Soc.* 3 (1822): 421–429; John Damper Parks, "Upon the Proper Management of Plants during Their Voyage from China to England," *Trans. Hort. Soc.* 7 (1830): 396–399; Main, "Reminiscences of a Voyage to and from China," 337–339; John Lindley, *Theory and Practice of Horticulture* (London: Longman, Orme, Brown, Green, and Longmans, 1855), 245–261; idem, "Instructions for Packing Living Plants in Foreign Countries, Especially within the Tropics; and Directions for Their Treatment during the Voyage to Europe," *Trans. Hort. Soc.* 5 (1824): 192–200; Kenneth Lemmon, *The Golden Age of Plant Hunters* (London: Aldine Press, 1968), 121–127.

153. On the Wardian case, see, for example, Allen, *The Naturalist in Britain,* 118–120. In China, Robert Fortune, the famous tea hunter, pioneered in using Wardian cases to transport plants. Fletcher, *The Story of the Royal Horticultural Society,* 149–150. The development of the overland route via the Red Sea about the middle of the century also greatly shortened the distance and time of traveling between Europe and China.

154. BL. BM. Add. MS. 33981. 233.

155. Kerr to Banks, 24 February 1806, BL. BM. Add. MS. 33981. 227–228.

156. BL. BM. Add. MS. 33981. 227–28; BM. Add. 33981. 234–235; DTC, vol. 17, f. 38. It appears that the Kew staff and Kerr did not like Au Hey, and Kew did not continue the practice of hiring a Chinese gardener to accompany the plants home. Desmond, *Kew: The History of the Royal Botanic Gardens,* 122; Coats, *The Plant Hunters,* 99. The East Indiamen routinely recruited large numbers of

Chinese as laborers and seamen for the voyages, and some of them might have been assigned to look after the plants on shipboard. During the Napoleonic Wars, the homebound East Indiaman *Warren Hastings* was captured by a French frigate because its Chinese crew, no fewer than forty men, had stayed in Canton and the ship was seriously undermanned. Sutton, *Lords of the East*, 38.

157. *Gardener's Magazine* 9 (1833): 474–475.

158. Charles Millet to W. J. Hooker, 6 May 1833, Kew: Directors Correspondence. 53 (87).

159. Ibid. See also *Gardener's Magazine* 9 (1833): 474–475.

160. Livingstone, "Observations on the Difficulties which have existed in the Transportation of Plants from China to England," 427. His estimation included even the most inexperienced captains or sailors who picked up plants in Canton and tried to ship them to Britain. Most of these attempts of course failed.

161. *The Botanical Register* 5 (1819), pl. 397.

2. Art, Commerce, and Natural History

1. Ann Secord, "Botany on a Plate: Pleasure and the Power of Pictures in Promoting Early Nineteenth-Century Scientific Knowledge," *Isis* 93 (2002): 28–57.

2. The literature on Chinese export painting is extensive and written mostly by connoisseurs and museum curators. Carl L. Crossman, *The China Trade* (Princeton N. J.: Pyne Press, 1972); Margaret Jourdain and R. Soame Jenyns, *Chinese Export Art in the Eighteenth Century* (Feltham, England: Spring Books, 1967 [1950]); Craig Clunas, *Chinese Export Watercolours* (London: Victoria and Albert Museum, 1984); Patrick Connor, *The China Trade, 1600–1800* (Brighton, England: The Royal Pavillion, 1986); Craig Clunas, ed., *Chinese Export Art and Design* (London: Victoria and Albert Museum, 1987); David S. Howard, *A Tale of Three Cities: Canton, Shanghai & Hong Kong: Three Centuries of Sino-British Trade in the Decorative Art* (London: Sotheby's, 1997); idem, *The Choice of the Private Trader: The Private Market in Chinese Export Porcelain Illustrated from the Hodroff Collection* (London: Zwemmer, 1994); Hong Kong Museum of Art, *Late Qing China Trade Paintings* (Hong Kong: The Urban Council, 1982); Hong Kong Museum of Art, *Gateways to China: Trading Ports of the Eighteenth and Nineteenth Centuries, Presented by the Hong Kong Museum of Art* (Hong Kong: The Urban Council, 1987); idem, *Views of the Pearl River Delta: Macau, Canton, and Hong Kong* (Hong Kong and Salem: The Hong Kong Museum of Art and the Peabody Essex Museum, 1997). I have also used the catalogs published by Martyn Gregory, a major dealer in Chinese export art in London. For example, *Trade Winds to China* (1987); *The China Trade* (1996); *Tingqua's China* (1986); and *Artist of the China Coast* (1991).

3. Like Chinese artists for export painting, the Indian artists also developed a hybrid style, called the Company School, and their natural history drawings for the British displayed the peculiar style. See Mildred Archer, *Natural History Drawings in the India Office Library* (London: HMSO, 1962), 53–57.

4. The literature on visual representation in science is sizable. Recent works on the subject include Brian Braigrie, ed., *Picturing Knowledge: Historical and Philosophical Problems Concerning the Use of Art in Science* (Toronto: University of Toronto Press, 1996); Barbara Stafford, *Artful Science: Enlightenment, Entertainment, and the*

Eclipse of Visual Education (Cambridge, Mass.: MIT Press, 1994); Caroline Jones and Peter Galison, eds., *Picturing Science, Producing Art* (New York: Routledge, 1998); Renato G. Mazzolini, ed., *Non-Verbal Communication in Science prior to 1900* (Firenze, Italy: Leo S. Olschki, 1993); Gregg Mitman, *Reel Nature: America's Romance with Wildlife on Film* (Cambridge, Mass.: Harvard University Press, 1999); Jennifer Tucker, "Photography as Witness, Detective, and Impostor: Visual Representation in Victorian Science," in *Victorian Science in Context*, ed. Benard Lightman (Chicago: University of Chicago Press, 1997), 378–408; Daniel Fox and Christopher Lawrence, eds., *Photographing Medicine: Images and Power in Britain and America in Science and Medicine between the Eighteenth and Twentieth Centuries* (New York: Greenwood Press, 1988); Greg Myers, "Every Picture Tells a Story: Illustrations in E. O. Wilson's Sociobiology," in Michael Lynch and Steve Woolgar, eds., *Representation in Scientific Practice* (Cambridge, Mass.: MIT Press, 1990), 231–266; Nicolaas A. Rupke, "'The End of History' in the Early Picturing of Geological Time," *History of Science* 36 (1998): 61–90; Martin Rudwick, "A Visual Language for Geology," *History of Science* 14 (1976): 149–195 and his *Scenes from Deep Time: Early Pictorial Representations of the Prehistoric World* (Chicago: University of Chicago Press, 1992); Brian Ford, *Images of Science: A History of Scientific Illustration* (London: British Library, 1992); Thomas Hankins, "Blood, Dirt, and Nomograms: A Particular History of Graphs," *Isis* 90 (1999): 50–80; "Seeing Science," a special issue of *Representations*, no. 40 (Fall 1992); and the special issue on science and the visual of *British Journal of the History of Science* 31, no. 2 (June 1998).

5. On natural history illustrations, see, for example, Brian Dolan, "Pedagogy through Print: James Sowerby, John Mawe and the Problem of Colour in Early Nineteenth-Century Natural History Illustration," *British Journal of the History of Science* 31 (1998): 275–304; Martin Kemp, "Taking It on Trust: Form and Meaning in Naturalistic Representation," *Archives of Natural History* 17 (1990): 127–188; Sachiko Kusukawa, "Leonhart Fuchs on the Importance of Pictures," *Journal of the History of Ideas* 58 (1997): 403–427; Ann Shelby Blum, *Picturing Nature: American Nineteenth-Century Zoological Illustration* (Princeton: Princeton University Press, 1993); David Knight, *Zoological Illustration: An Essay towards a History of Printed Zoological Pictures* (Folkstone, England: Dawson, 1977); S. Peter Dance, *The Art of Natural History: Animal Illustrators and Their Work* (Woodstock, N.Y.: Overlook Press, 1978); Wilfrid Blunt and William Stearn, *The Art of Botanical Illustration*, new ed. (Woodbridge, England: Antique Collectors' Club, 1994); Ray Desmond, *Wonders of Creation: Natural History Drawings in the British Library* (London: The British Library, 1986).

6. On Merian: Londa Schiebinger, *The Mind Has No Sex?: Women in the Origins of Modern Science* (Cambridge, Mass.: Harvard University Press, 1989), 68–79; Natalie Zemon Davis, *Women on the Margins: Three Seventeenth-Century Lives* (Cambridge, Mass.: Harvard University Press, 1995), 140–202; Tomomi Kinukawa, "Art Competes with Nature: Maria Sibylla Merian (1647–1717) and the Culture of Natural History" (Ph.D. diss., University of Wisconsin-Madison, 2001). On Ehret: Blunt and Stearn, *The Art of Botanical Illustration*, 159–174. On Audubon: Blum, *Picturing Nature*, ch. 3; Robert Henry Welker, *Birds and Men: American Birds in Science, Art, Literature and Conservation, 1800–1900* (Cambridge, Mass.: Harvard University Press, 1955), ch. 6.

7. In Edgerton's discussion, the Chinese illustrations serve simply as a conve-

nient example of products of a visual culture alien to the one developed in Renaissance Europe. They are tangential to his argument. Samuel Edgerton, *The Heritage of Giotto's Geometry: Art and Science on the Eve of the Scientific Revolution* (Ithaca: Cornell University Press, 1991), ch. 8. Compare Michael Mahoney, "Diagrams and Dynamics: Mathematical Perspectives on Edgerton's Thesis," in J. W. Shirley and F. D. Hoeniger, eds., *Science and the Arts in the Renaissance* (Washington, D.C.: Folger Shakespeare Library, 1985), 198–220. On the general subject of "realism" and conventions of visual representation, E. H. Gombrich's classic study remains valuable. See his *Art and Illusion: A Study in the Psychology of Pictorial Representation* (Princeton: Princeton University Press, 1960).

8. For biographic information about John Reeves, see *Gardeners Chronicle*, 29 March 1856, p. 212; Bretschneider, *History*, 256–263; *Dictionary of National Biography*, vol. 16, 859.

9. They were originally letters to the editors of the journals. For example, J. Reeves, "Account of Some of the Articles of the Materia Medica Employed by the Chinese," *Transactions of the Medico-Botanical Society of London* 1 (June 1828): 24–27; idem, "A Comet Observed in China, in May, 1820," *The Indo-Chinese Gleaner* 2 (October 1820): 436–439; *Gardener's Magazine* 11 (1835): 112.

10. R. Morrison, *Dictionary of the Chinese Language* (Macao: Printed at the Honorable East India Company's Press, 1819), part II, p. 1063.

11. Reeves's letters (not dated or numbered) in RHS. 69D15.

12. Bretschneider, *History*, 263–266.

13. For example, DTC, vol. 18, 194; vol. 19, 80.

14. Bretschneider, *History*, 266–268.

15. Hunter, *Bits of Old China*, 73–78; *Chinese Repository* 11 (January–December 1842): 59–60.

16. Hunter, *Bits of Old China*, 74.

17. *Chinese Repository* 11:59–60.

18. Vachell to John Henslow, 30 January 1831, CUL. Add. 8176. No. 190. A less lucky visitor saw only two hundred, but that was probably in a different year. See S. W. Williams, "Recollections of China prior to 1840," *JNCB*, n.s., 8 (1873): 8.

19. George Bennett, *Wanderings in New South Wales, Batavia, Pedir Coast, Singapore, and China; being the Journey of a Naturalist in those Countries, during 1832, 1833, and 1834*, vol. 2 (London: Richard Bentley, 1834), 36–80.

20. Ibid., 57.

21. On Reeves's drawings, see Patrick M. Synge, "Chinese Flower Paintings: An Important Purchase by the Royal Horticultural Society," *Journal of the Royal Horticultural Society* 78 (1953): 209–213; P. J. P. Whitehead and P. I. Edwards, *Chinese Natural History Drawings Selected from the Reeves Collection in the British Museum (Natural History)* (London: The Trustees of the British Museum, 1974).

22. James Anderson, *Letters to Sir Joseph Banks, Baronet, President of the Royal Society, on the Subject of Cochineal Insects, Discovered at Madras, &c. &c.* (Madras: Printed by Charles Ford, 1788), 24. The letter was from James Kincaid to James Anderson, 10 November 1787.

23. J. M. [James Main], letter to editor of *Gardener's Magazine* 2 (1827): 424.

24. James Main, "Reminiscences of a Voyage to and from Canton, in the Year 1792-3-4," *Horticultural Register* 5 (1836): 148.

25. Bretschneider, *History*, 33. Desmond, *Wonders of Creation*, 128.

26. Desmond, *Wonders of Creation*. The drawings are now in the Botany Library at the Natural History Museum of London.

27. OIOC. MSS. EUR. D562/16. "Extract Secret Letter from Canton dated the 29th January 1804."

28. Main, "Reminiscences of a Voyage to and from China," 148.

29. Joseph Banks, "Hints on the Subject of Gardening, suggested to the Gentlemen who attend the Embassy to China," The Linnean Society. MSS, f. 4; DTC, vol. 14, f. 66; BL. BM. Add. MS. 33978. 276–279; Add. MS. 33979. 14; Add. MS. 22979. 15–16.

30. Clunas, ed., *Chinese Export Art and Design* provides a brief survey of the different kinds of artwork produced in Old Canton. It should be noted that Cantonese artisans also produced hybrid works of art and curiosity, notably clocks, for domestic consumption. Catherine Pagani, *"Eastern Magnificence & European Ingenuity": Clocks of Late Imperial China* (Ann Arbor: University of Michigan Press, 2001), chs. 2 and 3.

31. James Cahill and Michael Sullivan have argued that certain literati artists had already absorbed Western pictorial techniques as early as the seventeenth century. They might have learned the techniques from the books and prints by the missionaries. The examples they cited are few, however. James Cahill, *The Compelling Image: Nature and Style in Seventeenth-Century Chinese Painting* (Cambridge, Mass.: Harvard University Press, 1982), chs. 1 and 3; Michael Sullivan, *The Meeting of Eastern and Western Art*, rev. ed. (Berkeley: University of California, 1989), ch. 2.

32. Howard, *The Choice of the Private Trader*, 26.

33. Peter Quennell, ed., *The Prodigal Rake: Memoirs of William Hickey* (New York: E. P. Dutton, 1962), 140.

34. On the Dutch designs, see Jessica Rawson, *The British Museum Book of Chinese Art* (London: Thames and Hudson, 1992), 279–280; Clunas, ed., *Chinese Export Art and Design*, 64.

35. Josiah Quincy, ed., *The Journal of Major Samuel Shaw, the First American Consul at Canton, with a Life of the Author, by Josiah Quincy* (Boston, Mass.: Wm. Crosby and H. P. Nichols, 1847), 198–199.

36. For example, *China's Influence on American Culture in the 18th and 19th Centuries*, 36–37; Howard, *A Tale of Three Cities*, 119; Clunas, ed., *Chinese Export Art and Design*, 64; Arlene M. Palmer, *A Winterthur Guide to Chinese Export Porcelain* (New York: Rutledge Book, 1976), 85, 131.

37. *The Chinese Repository*: 4 (May 1835–April 1836), 291.

38. Quincy, *The Journals of Major Samuel Shaw*, 198.

39. Palmer, *A Winterthur Guide to Chinese Export Porcelain*, 23.

40. Downing, *The Fan-Qui in China*, 83–84; Palmer, *A Winterthur Guide to Chinese Export Porcelain*, 23.

41. Downing, *The Fan-Qui in China*, 96–99.

42. Clunas, *Chinese Export Watercolours*, 10. Chinese painters elsewhere were also quick to take advantage of this kind of "postcard industry," so to speak. During the Second Opium War, the British and French armies occupied Tianjin, northern China, and the local painters seized the opportunity of drawing portraits and caricatures of these odd-looking foreign soldiers, some in red (the English) and others in blue (the French). Most of the pictures were about the market scenes and ev-

eryday life of the alien soldiers in the Chinese city. Mrs. Muter, wife of Lieut.-Colonel D. D. Muter, *Travels and Adventures of an Officer's Wife in India, China, and New Zealand*, vol. 2 (London: Hurst and Blackett, 1864), 73–75.

43. Grossman, *The China Trade*, ch. 2; Patrick Connor, *George Chinnery (1774–1852): Artist of India and the China Coast* (Woodbridge, England: Antique Collectors' Club, 1993), 263–268; Sander L. Gilman, "Lam Qua and the Development of a Westernized Medical Iconography in China," *Medical History* 30 (1986): 57–69; Larissa N. Heinrich, "Handmaids to the Gospel: Lam Qua's Medical Portraiture," in Lydia Liu, ed., *Tokens of Exchange: The Problem of Translation in Global Circulations* (Durham: Duke University Press, 1999), 239–275.

44. *The Canton Register*, 8 December 1835. C. Toogood Downing, who was himself a sketcher, described in detail Lamqua's studio. Downing, *The Fan-Qui in China*, vol. 2, 90–117.

45. *Tingqua's China*. Exhibition Catalogue (London: Martyn Gregory, 1986), no. 1.

46. Joseph D. P. Ting's introduction to Hong Kong Museum of Art, *Late Qing China Trade Paintings*.

47. DTC, vol. 18, 192–198. We also know that some of Tongqua's paintings of Canton were made into panoramas for exhibition in Britain. *The Canton Register*, vol. 3, no. 50, 18 August 1838.

48. Whitehead and Edwards, *Chinese Natural History Drawings*, 22.

49. The four artists were commissioned to draw fish and probably other subjects as well. NHML: John Reeves. Z.88.ff.R.

50. DTC, vol. 14, f. 66.

51. Bennett, *Wanderings in New South Wales*, 61.

52. *Proc. Zool. Soc.* (1862): 220.

53. This contrast is deliberately simplified. The two traditions were practiced by socially distinct groups, and there were other traditions of Chinese painting that were accepted by the literati during this period. But further nuances are not necessary for our purpose. On the literati painting, see, for example, Lin Mu, *Min-Qing wenren hua xinchao* (Shanghai: Shanghai renmin meishu chubanshe, 1991). James Cahill's pioneering study of the practice of the literatus-painter in late imperial China is essential reading. James Cahill, *The Painter's Practice: How Artists Lived and Worked in Traditional China* (New York: Columbia University Press, 1994).

54. André Georges Haudricourt and Georges Métailié, "De l'illustration botanique en Chine," *Études chinoises* 8, nos. 1–2 (1994): 381–416; Richard Rudolph, "Illustrated Botanical Works in China and Japan," in Thomas R. Buckman, ed., *Bibliography and Natural History* (Lawrence: University of Kansas Press, 1966), 103–120; Zhongguo zhiwu xuehui, ed., *Zhongguo zhiwu xueshi* (Beijing: Kexue chubanshe, 1994), ch. 17.

55. Later in the nineteenth century, when British naturalists had more contact with Chinese works on plants and animals, they would consider the illustrations in Wu Qijun's pictorial dictionary of plants, *Zhiwu mingshi tukao* (1848), of high quality. More on Western naturalists' perceptions of the illustrations in Chinese works on animals and plants in Chapter 4.

56. Reeves was not one of the language students at the East India Company, but he worked with Robert Morrison, the sinologist, on Chinese natural history

and *materia medica*. See Chapter 1. Also, he possessed a copy of *Bencao gangmu*, which he was able to consult, probably with the assistance of Morrison's dictionary. See his letter to George Bentham, 3 March 1835, Kew: Bentham Correspondence. 7 (3311).

57. These conventions were obviously learned by apprenticeship. Some of the conventions were not dissimilar to those in the literati painting, which were much better documented. The popular manual of painting *Jiezi yuan hua pu* (1679–1701), for example, laid out the principles of representing the subjects, be they flowers, rocks, trees, birds, or human figures. The work is available in English translation. Wang Kai, *The Mustard Seed Garden Manual of Painting* (Princeton: Princeton University Press, 1977). Paul Hulton and Lawrence Smith, *Flowers in Art from East and West* (London: The Trustees of the British Museum, 1979) includes some introductory essays on different kinds of flower painting in China.

58. Whitehead and Edwards, *Chinese Natural History Drawings*, 21; Desmond, *Wonders of Creation*, frontispiece and pl. 49. The entomologist J. O. Westwood stated in his preface to Edward Donovan, *Natural History of the Insects of China*, new ed. (London: Henry G. Bohn, 1842) that "numerous beautiful drawings of insects upon rice paper [were] brought to Europe in great quantities." He also noted that "[m]any of these figures are, however, evidently fictitious, although some are occasionally found accurately correct and most elaborately pencilled."

59. It was mentioned in an obituary on Reeves that the drawings were "executed in his own house." *Gardeners Chronicle*, 29 March 1856, p. 212. But it can be gathered from Reeves's notebook on his fish drawings that the artists brought the finished drawings to him for examination instead of working on them at Reeves's place. See NHML: John Reeves. Z. 88. ff. R.

60. The issue of scientific illustrations and objectivity is discussed in Lorraine Daston and Peter Galison, "The Image of Objectivity," *Representations*, no. 40 (Fall 1992): 81–128.

61. William Jardine, *The Naturalist's Library*, 40 vols. (Edinburgh: W. H. Lizars, 1843).

62. For example, NHML: Reeves. Botanical Drawings. 138, 960. Unless the artists were the same ones employed by James Main, who visited Canton from 1792 to 1794, or William Kerr, who stayed in Canton and Macao between 1803 and 1812. In any case, they had to learn from British naturalists in China to depict the sexual system of the plant, the cross section of the fruit, and other ways of representing a plant for scientific purpose.

63. For example, NHML: Reeves. Zoological Drawings. L. S. 81; L. S. 76.

64. He attempted to identify some of the subjects of his fish drawings without much success. See his notebook on fish drawings. In fact, he never classified or named a plant or an animal on his own. But according to the late Peter Whitehead, who knew more about Reeves's fish drawings than anybody, Reeves showed some facility for sketching and an acquaintance with the Chinese fishes. See Whitehead and Edwards, *Chinese Natural History Drawings*, 22.

65. On the stylistic elements of Chinese export painting of plants and animals, see Gill Saunders, *Picturing Plants: An Analytic History of Botanical Illustration* (Berkeley: University of California Press, 1995), 80–81; Clunas, *Chinese Export Watercolours*, 84–89; Archer, *Natural History Drawings in the India Office Library*, 61.

66. NHML: Reeves. Botanical Drawings. 407.

67. NHML: Reeves. Zoological Drawings. 62. In her discussion of the botanical illustrations executed by Indian artists in collaboration with, or under the supervision of, British naturalists, Beth Fowkes Tobin interprets the stylistic traits of Indian painting in the drawings as colonial resistance against imperial domination. Of course, the relationships of Britain to India and China were very different, but I am not sure that it makes much sense to fit everything into a rigid, static, and undifferentiating framework of postcolonial criticism of "imperialism" without taking into consideration historical actors, motivations, local context, and time period—even with regard to the case of India. It is even less meaningful to find resistance everywhere. Beth Fowkes Tobin, *Picturing Imperial Power: Colonial Subjects in Eighteenth-Century British Painting* (Durham: Duke University Press, 1999), 174–201.

68. Audubon's early drawings for *Ornithological Biography* met with criticisms because they did not have scientific descriptions, but he soon sought help from the ornithologist William MacGillivray to remedy this deficiency. Blum, *Picturing Nature*, 112–113.

69. The Canton Factory did have a library, a large portion of which was Robert Morrison's collections. The library, of course, was inadequate for taxonomic work. The "British Museum in China" was proposed during Reeves's last years in China but was soon abandoned with the end of the East India Company's monopoly of the British China trade. See Chapter 1.

70. Joseph Sabine, "Further Account of Chinese Chrysanthemums; with Descriptions of several New Varieties," *Trans. Hort. Soc.*, sr. 1, 5 (1824): 412–428. On p. 425.

71. Ibid., 425.

72. On the flexibility of networks of scientific information in the early modern period, see David Lux and Harold Cook, "Closed Circles or Open Networks?: Communicating at a Distance during the Scientific Revolutions," *History of Science* 36 (1998): 179–211.

73. Commenting on some perished specimens of fish from China, Albert Günter, then Assistant Keeper, Zoological Department at the British Museum, said, "Only too frequently specimens are sent off by collectors before they are thoroughly saturated with strong spirits; and the inevitable consequences are that they are softened by internal decomposition and knocked to pieces by the rough treatment to which packages are subjected during a long journey. All specimens collected in a hot climate and placed in spirits ought to be retained by the collectors for at least four weeks before they are sent off, and the spirits changed two or three times." *Annals and Magazine of Natural History*, sr. 4, 12 (1873): 239–250. This complaint was made when the duration of a journey from China to Britain had been reduced to little more than a month because of the improved transport and the development of the overland route.

74. P. J. P. Whitehead, "The Elopoid and Clupeoid Fishes in Richardson's 'Ichthology [sic] of the Seas of China and Japan 1846," *The Bulletin of the British Museum (Natural History), Zoological Series* 14, no. 2 (1966): 17–52. For von Siebold's natural historical research in Japan, see L. B. Holthuis and T. Sakai, *Ph. F. von Siebold and Fauna Japonica: A History of Early Japanese Zoology* (Tokyo: Academic Press of Japan, 1970); John Z. Bowers, *Western Medical Pioneers in Feudal Japan* (Baltimore: Johns Hopkins University Press, 1970), chs. 4 and 5; J. Mac Lean,

"Von Siebold and the Importation of Japanese Plants into Europe *via* the Netherlands," *Japanese Studies in the History of Science*, no. 17 (1978): 43–79.

75. See Reeves's notebook on fish drawings, NHML: John Reeves. Z.88.ff.R. Reeves recorded in the notebook the subjects (sometimes with species names and their local Chinese names), the transactions with the artists, and other information, such as if the fish was common in Canton or if they tasted good. For example, f. 1, "Sheet 8 . . . looks so like an artificial painting that I procured another specimen and had painted by another man—which is added here. The Colours are very fleeting and so change after the fish is dead."

76. John Richardson, "Report on the Ichthyology of the Seas of China and Japan," *Report of the 15th Meeting of the British Association of the Advancement of Science* (1845): 187–320. On p. 188.

77. Ibid., 188.

78. Ibid., 187.

79. It was a common practice among British naturalists in India to employ native artists to draw natural history illustrations. See Mildred Archer, "Indian Paintings for British Naturalists," *The Geographic Magazine* 28 (1955): 220–230.

80. Its importance was overshadowed by the ichthyology volumes of P. F. von Siebold's *Fauna Japonica* (1833–50), which appeared about the same time. However, not only did Richardson's paper supplement *Fauna Japonica*, but it provided critical data for the study of the geographic distribution of fish in the eastern seas.

81. Kerr: G. T. Staunton's letters to the Court Directors of the East India Company, 29 January 1804, 23 January 1805, 26 February 1806, OIOC. MSS. EUR. D562/16.

82. For example, Robert Fortune to John Lindley, 22 March 1844, RHS. Robert Fortune Correspondence. 69D16.

83. Mildred Archer and John Bastin, *The Raffles Drawings in the India Office Library London* (Oxford: Oxford University Press, 1978), 8; Archer, *Natural History Drawings in the India Office Library*, 61.

84. Theodore Cantor's drawings are kept in OIOC. NHD 8. Nos. 1151–1292. Robert Fortune to John Lindley, 31 January 1844, RHS. Robert Fortune. Correspondence. 69D16. Hance's drawings are at Kew Gardens. The painters working with Cantor and Fortune were on the island of Chusan (Zhoushan) and had not had much contact with Western art. However, they were obviously well trained in the strong tradition of Chinese drawings of plants and animals. It may be worth noting that the Russian physician Alexander Tatarinov in Beijing also employed a Chinese artist to draw plants. The drawings were said to be "executed from nature" and "carefully colored," and it also showed "the botanical details of each specimen." Bretschneider, "Botanicon Sinicum," *JNCB*, n.s., 16 (1881): 123.

3. Science and Informal Empire

1. Randall A. Dodgen, *Controlling the Dragon: Confucian Engineers and the Yellow River in Late Imperial China* (Honolulu: University of Hawaii Press, 2001).

2. Standard surveys of Sino-Western relations in the nineteenth century include Jonathan Spence, *The Search for Modern China* (New York: Norton, 1990); Immanuel C. Y. Hsu, *The Rise of Modern China*, 5th ed. (New York: Oxford University, 1995); John K. Fairbank et al., ed., *The Cambridge History of China*, vol. 10, parts I and II (Cambridge, England: Cambridge University Press, 1978–1980).

Hosea Ballou Morse's *The International Relations of the Chinese Empire*, 3 vols. (Taipei: Wenshing shudian, 1963 [1911–17]) remains a convenient source of factual information. The literature on the Opium War is legion. See Peter Ward Fay, *The Opium War, 1840–42: Barbarians in the Celestial Empire in the Early Part of the Nineteenth Century and the War by Which They Forced Her Gates Ajar* (Chapel Hill: University of North Carolina Press, 1975) and James M. Polachek, *The Inner Opium War* (Cambridge, Mass.: Harvard University Press, 1992) for two different perspectives. Several specialized books also provide useful background information for the present chapter. John K. Fairbank, ed., *The Chinese World Order: Traditional China's Foreign Relations* (Cambridge, Mass.: Harvard University Press, 1968); Immanuel Hsu, *China's Entrance into the Family of Nations: The Diplomatic Phase, 1858–1880* (Cambridge, Mass.: Harvard University Press, 1960); Masataka Banno, *China and the West, 1858–1861: The Origins of the Tsungli Yamen* (Cambridge, Mass.: Harvard University Press, 1964); John K. Fairbank, *Trade and Diplomacy on the Chinese Coast: The Opening of the Treaty Ports, 1842–1854*, 2 vols. (Cambridge, Mass.: Harvard University Press, 1953); Frederic Wakeman, Jr., *Strangers at the Gate: Social Disorder in South China, 1839–1861* (Berkeley: University of California Press, 1966).

3. W. H. Medhurst, *The Foreigner in Far Cathay* (London: Edward Stanford, 1872) discusses the three major groups of Western residents in China, excluding the working class, and their life in China. See also N. B. Dennys, *The Treaty Ports of China and Japan: A Complete Guide to the Open Ports of Those Countries, together with Peking, Yedo, Hongkong and Macao: Forming a Guide Book & Vade Mecum for Travellers, Merchants, and Residents in General* (San Francisco: Chinese Materials Center, 1977 [1867]). For secondary accounts, see, for example, Wellington Chan, *Merchants, Mandarins and Modern Enterprise in Late Ch'ing China* (Cambridge, Mass.: Harvard University Press, 1977); Liu Kwang-Ching, *Anglo-American Steamship Rivalry in China, 1862–1874* (Cambridge, Mass.: Harvard University Press, 1962); Yen-p'ing Hao, *The Commercial Revolution in Nineteenth-Century China: The Rise of Sino-Western Mercantile Capitalism* (Berkeley: University of California, 1986); P. D. Coates, *The China Consuls* (Oxford: Oxford University Press, 1988); Robert Bickers, "Shanghailanders: The Formation and Identity of the British Settler Community in Shanghai, 1842–1937," *Past and Present* 159 (1998): 161–211.

4. For biographic information of Robert Fortune, E. H. Wilson, and George Forrest, see E. H. M. Cox, *Plant Hunting in China: A History of Botanical Exploration in China and the Tibetan Marches* (London: Collins, 1945), 70–96, 136–169. Alice Coats, *The Plant Hunters* (New York: McGraw-Hill, 1970), 101–110, 118–127. Bretschneider gives a detailed account of Fortune's travels in China and Japan in his *History*, 403–518.

5. For example, Lucile Brockway, *Science and Colonial Expansion: The Role of the British Royal Botanical Gardens* (New York: Academic Press, 1979); Donal P. McCracken, *Gardens of Empire: Botanical Institutions of the Victorian British Empire* (London: Leicester University Press, 1997); Richard Grove, *Green Imperialism: Colonial Expansion, Tropical Island Edens and the Origins of Environmentalism, 1600–1800* (Cambridge, England: Cambridge University Press, 1995); James E. McClellan, III, *Colonialism and Science: Saint Domingue in the Old Regime* (Baltimore: Johns Hopkins University Press, 1992), ch. 9; Richard Drayton, *Nature's Government: Science, Imperial Britain, and the "Improvement" of the World* (New Haven: Yale University Press, 2000); Daniel Headrick, *The Tentacles of Progress: Techno-*

logical Transfer in the Age of Imperialism, 1850–1940 (Oxford: Oxford University Press, 1988), 209–258; William Kelleher Storey, *Science and Power in Colonial Mauritius* (Rochester: University of Rochester Press, 1997).

6. For the early history of Hong Kong, see Frank Welsh, *A History of Hong Kong* (London: HarperCollins, 1993), 101–154; G. B. Endacott, *A History of Hong Kong*, 2nd ed. (Hong Kong: Oxford University Press, 1973), 14–34, and, from a different perspective, Jung-fang Tsai, *Hong Kong in Chinese History: Community and Social Unrest in the British Colony, 1842–1913* (New York: Columbia University Press, 1993), 17–35.

7. See, for example, Coates, *The China Consuls;* Stanley Wright, *Hart and the Chinese Customs* (Belfast: Mullan, 1950); K. S. Latourette, *A History of Christian Missions in China* (Taipei: Chengwen Publishing Co., 1966 [1929]); Paul Cohen, "Christian Missions and Their Impact to 1900," *The Cambridge History of China*, vol. 10, part I, 543–590; D. A. Griffiths and S. P. Lau, "The Hong Kong Botanical Gardens: A Historical Overview," *Journal of the Hong Kong Branch of the Royal Asiatic Society* 26 (1986): 55–77.

8. Jürgen Ostenhammel, "Britain and China," in Andrew Porter, ed., *The Oxford History of the British Empire: The Nineteenth Century* (Oxford: Oxford University Press, 1999), 146–169, on p. 156.

9. On information technologies and imperialism/colonialism, see, in particular, C. A. Bayly's *Empire of Information: Intelligence Gathering and Social Communication in India, 1780–1870* (New York: Cambridge University Press, 1996), which pioneers the study of information gathering, social communication, and imperialism. See also Matthew H. Edney, *Mapping an Empire: The Geographical Construction of British India, 1765–1843* (Chicago: University of Chicago Press, 1997); Richard Drayton, "Knowledge and Empire," in P. J. Marshall, ed., *Oxford History of the British Empire: Eighteenth Century* (Oxford: Oxford University Press, 1998), 321–352; David Ludden, "Orientalist Empiricism: Transformations of Colonial Knowledge," in Carol A. Breckenridge and Peter van der Veer, eds., *Orientalism and the Postcolonial Predicament* (Philadelphia: University of Pennsylvania Press, 1993), 250–278; Arjun Appadurai, "Number in the Colonial Imagination," ibid., 314–339; Deepak Kumar, *Science and the Raj, 1857–1905* (Delhi: Oxford University Press, 1997), ch. 3; Marika Vicziany, "Imperialism, Botany and Statistics in Early Nineteenth-Century India: The Surveys of Francis Buchanan (1762–1829)," *Modern Asian Studies* 20 (1986): 625–660; Daniel Headrick notices the issue of imperial communications, though his focus is very different from mine. Daniel Headrick, *The Tools of Empire: Technology and European Imperialism in the Nineteenth Century* (New York: Oxford University Press, 1981), part III; idem, *The Tentacles of Progress* (cited in note 5).

10. Ostenhammel, "Britain and China," 156.

11. Ray Desmond, *The European Discovery of the Indian Flora* (New York: Oxford University Press, 1992), 220–230; Headrick, *The Tools of Empire*, 58–79; Drayton, *Nature's Government*, chs. 6 and 7.

12. Keith Sinclair, ed., *A Soldier's View of Empire: The Reminiscences of James Bodell, 1831–92* (London: The Bodley Head, 1982), 58. On the high death rate of Europeans in the tropics, see Philip D. Curtin, *Death by Migration: Europe's Encounter with the Tropical World in the Nineteenth Century* (New York: Cambridge University Press, 1989).

Notes to Pages 65–67   195

   13. Grove, *Green Imperialism*, chs. 5–8.
   14. Kew: Misc. Reports. 4.41. Hong Kong. Botanic Gardens. 1870–1915, ff. 2–19.
   15. Kew: Misc. Reports. 4.41. Hong Kong. Botanical Gardens, 1870–1915, ff. 28–30; Public Records Office (Hong Kong). CO 129/189, 129/190.
   16. Edward Kynaston, *A Man on Edge: A Life of Baron Sir Ferdinand von Mueller* (London: Allen Lane, 1981), 269–273, 280–292.
   17. W. W. Perry to Thiselton-Dyer, Kew: Chinese and Japanese Letters, 151 (850); Kew: Misc. Reports. 4.41. Hong Kong. Botanic Gardens. 1870–1915, ff. 67–68.
   18. Kew: Chinese and Japanese Letters, 150 (546), (550), (551), (850).
   19. Kew: Misc. Reports. 4.41. Hong Kong. Botanic Gardens. 1870–1915, ff. 40–43. The botanist Henry Hance, who was involved in the planning of the Gardens, knew this very well, and he urged the decision makers to choose "the advancement of science" over "promenading" as the primary objective of the development of the Gardens. See Berthold Seemann, *Narrative of the Voyage of H.M.S. Herald during the Years 1845–51, under the Command of Captain Henry Kellett . . . ,* vol. 2 (London: Reeve and Company, 1853), 227.
   20. Hance to Hooker, Kew: Chinese and Japanese Letters, 150 (551); Ford to J. D. Hooker, Kew: Chinese and Japanese Letters. 150 (198–199).
   21. Kew: Misc. Reports. 4.41. Hong Kong. Botanic Gardens. 1870–1915, ff. 108.
   22. Ibid.
   23. Ibid.
   24. Ibid.
   25. Kew: Misc. Reports. 4.41. Hong Kong. Botanical Gardens. 1870–1915, ff. 109–110.
   26. Ibid.
   27. Ibid.
   28. Ibid. (his emphasis). It is worth noting that Hooker and his scientific allies had just won a bitter controversy with the British government about the primary objective of Kew Gardens. The controversy was more or less parallel to, and must have had some impact on, the controversy concerning the Hong Kong Botanic Gardens. See Ray Desmond, *Kew: The History of the Royal Botanical Gardens* (London: Harvill Press, 1995), 223–250; R. M. MacLeod, "The Ayrton Incident: A Commentary on the Relations of Science and Government in England, 1870–73," in Arnold Thackray and Everett Mendelsohn, eds., *Science and Values* (New York: Humanities Press, 1974), 45–78.
   29. Kew: Misc. Reports. 4.41. Hong Kong Botanic Gardens. 1870–1915, f. 111.
   30. Public Records Office (Hong Kong). CO 129/217, 129/218.
   31. Commenting on the opposition to his scientific research, Ford admitted that "[it] is difficult to procure material benefits to Hong Kong from [botanical] expeditions . . . ," but he also regretted that "the people here do not feel the necessity of encouraging investigations for the sake of the acquirement of the knowledge only." Kew: Chinese and Japanese Letters. 150 (273–274). See also 150 (228–229), (247). Ford further complained to Thiselton-Dyer that "[t]here has always appeared to me a manifest jealousy on the part of some high officials of the Govt of the advancement of botanical science." In 150 (399–401).
   32. Hance's letters to J. D. Hooker, Kew: Chinese and Japanese Letters, 150 (546), (550).

33. See Ford's letters in Kew: Chinese and Japanese Letters, 150 (193), (239–240), (268–269); Kew: Misc. Reports. 4.41. Hong Kong. Botanical Gardens. 1870–1915, ff. 95–96.

34. Coates, *The China Consuls*, 201–203. Kew: Misc. Reports. 4.4. China. Hance, 2–4.

35. He had learned some botany before coming to China. *Dictionary of National Biography*, vol. 8, 1156.

36. Kew: Chinese and Japanese Letters, 150 (531).

37. Basic biographic facts of Hance can be found in *Gardeners Chronicle*, 14 August 1886, 218–219; *Hongkong Daily Press*, 26 and 28 June 1886; Bretschneider, *History*, 365–370, 632–652; *Dictionary of National Biography*, vol. 8, 1156–1157.

38. Kew: Bentham Correspondence. 5 (1768–69).

39. See, for example, George Bentham, *Florae Hongkongensis: A Description of the Flowering Plants and Ferns of the Island of Hongkong* (London: Lovell Reeve, 1861) and Henry Hance's "Supplement" (1871).

40. Comments on Hance's personality can be found in, for example, SOAS. Chaloner Alabaster Papers. MS 380451/2, entry under January 2; MS 380451/3, entry under June 1; E. H. Parker, "Henry Fletcher Hance," *JNCB*, n.s., 21 (1886): 309–313. After his death, Hance's friends tried to "correct an unfavourable impression" that many had of Hance's drinking problem. See Francis Forbes's letter to J. D. Hooker in Kew: Misc. Reports. 4.4. China. Hance, ff. 23–25, and Thomas Wade's letter in the same volume, f. 26.

41. His library contained several thousand volumes on a wide range of subjects. See the auction catalog of his library in Kew: Misc. Reports. 4.4. China. Hance, ff. 127–137.

42. About this time, Hance met several scientific friends, including John Champion, an army officer and entomologist, William Harland, surgeon and curator of the Hong Kong Museum, Berthold Carl Seemann, a botanist visiting Hong Kong, and later Robert Swinhoe, a consular officer and zoologist, and he also began correspondence with major botanists and botanical institutes in Europe. Kew: Bentham Correspondence. 5 (1768–1769), (1770–1772), (1773–1775), (1780–1786), (1790).

43. Bretschneider, *History*, 661–678. See also Hance's letters to Swinhoe in Kew: Kew Collectors IV. Richard Oldham (120), (122).

44. Kew: Chinese and Japanese Letters, 151(475).

45. See, for example, Bretschneider, *History*, 652–661; Hance's letters to J. D. Hooker, Kew: Chinese and Japanese Letters, 151 (493), (514–515), (519); Sampson to Thiselton-Dyer, 6 October 1886, Kew: Chinese and Japanese Letters, 151 (928); Hance to Swinhoe, 9 October 1864, Kew: Kew Collectors IV. Oldham (120). Sampson later became a teacher at a Chinese school and gave up botanical collecting. See Hance's letters to Daniel Hanbury in RPS: Hanbury Papers. P313 [3], [32].

46. Parker's respect and affections for "Old Hance" may be seen in his obituary of Hance in *JNCB*, n.s., 21 (1886): 309–313.

47. W. T. Thiselton-Dyer, "Historical Notes," in F. B. Forbes and W. B. Hemsley, *Index Florae Sinensis* (1905), v–xi. Details of the matter can be found in Kew: Misc. Reports. 4.4. China. Index Florae Sinensis, 1883–1905.

48. More on Bretschneider in Chapter 4. Bretschneider described Hance as "known to every educated European in China as a distinguished botanist." E.

Bretschneider, *On Chinese Silkworm Trees* (Peking: Privately printed, 1881), 1. He also urged Western residents in China to send plant specimens to Hance. See, for example, Bretschneider, *Notes on Some Botanical Questions Connected with the Export Trade of China* (Peking, 1880), 14. According to Hance, "[Bretschneider] often wrote to [him] for information, or for the determination of doubtful plants." Bretschneider paid Hance a visit in 1872, and Hance described the meeting as follows: "I never had such a continuous fire of questions on all sorts of botanical subjects put to me in my life. . . . [He] compelled me to produce hundreds of books and quotations." And Hance enjoyed the meeting very much. See RPS: Hanbury Papers. PH313[20], PH313[24].

49. The herbarium contained more than 22,400 different specimens and was sold to the British Museum (Natural History) after his death. Kew: Misc. Reports. 4.4. China. Hance, ff. 11, 75, 87.

50. Kew: Chinese and Japanese Letters, 151(531). There were few women naturalists in China, in part because the population of Westerners in China was overwhelmingly male, but also because, with rare exceptions, the social and political circumstances hardly permitted women to do fieldwork. The American missionary B. C. Henry's wife often accompanied her husband on journeys in the interior, and she had, in Henry Hance's words, "the charge and preparation, *ex officio*, of the botanical collections, a task for which the excellently dried specimens prove her to have been specially designed by natural selection." B. C. Henry, *Ling-Nam or Interior Views of Southern China* (London: S. W. Patridge & Co., 1886), 123–124. The wife of Horsea Morse, a Customs commissioner, probably also learned to handle specimens. See Kew: A. Henry. Letters to H. B. Morse, ff. 13, 19. Another example is T. H. Layton's wife, S. D. Layton. The plant used by the Chinese to make the so-called rice paper had long puzzled British botanists, and the problem received intense attention in the 1840s and 1850s when China became more accessible. There were wild guesses about the identity of the plant and the places where it could be found. In China, Sir John Bowring, Governor of Hong Kong, his son, and T. H. Layton, Consul at Amoy, all diligently investigated the matter. When Layton died, his wife took over the research and persuaded an "old brave Chinese admiral" to procure the plant for her. The mandarin ordered one to be brought from Taiwan, but unfortunately he, too, died before the mission could be completed. S. D. then sent two Chinese messengers to Taiwan to obtain some specimens and thus solved one of the long-standing mysteries in botany. It turned out to be a large pithy plant not uncommon in Taiwan. S. D. was apparently full of intellectual curiosity and took interest in Chinese *materia medica*. For the quest of the rice-paper plant, she also asked a Chinese member of the consulate at Amoy, "a most intelligent and learned man . . . who used much research in the old botanical works of his country," to draw up descriptions of the plant. S. D. Layton to W. J. Hooker, Kew: Directors Correspondence. 55 (192–192 E), (193–193B); W. J. Hooker, "On the Chinese Rice Paper," *Hooker's Journal of Botany* 4 (1852): 50–54. It is likely that some British women in China did flower painting, which was popular among educated Victorian women, although there were probably no peers of Marianne North, the famous Victorian traveler and botanical artist, in China. Hance's first wife was "once a very skilful flower-painter, but [had] long exchanged that accomplishment for the cultivation of babies." A talented woman, she was also a pianist. Hance to Daniel Hanbury. 5 February 1871, RPS: Hanbury Papers. P313[16];

Coates, *The China Consuls*, 90. On Victorian women naturalists, see Ann Shteir, *Cultivating Women, Cultivating Science: Flora's Daughters and Botany in England, 1760 to 1860* (Baltimore: Johns Hopkins University Press, 1996) and Barbara Gates, *Kindred Nature: Victorian and Edwardian Women Embrace the Living World* (Chicago: University of Chicago Press, 1998).

51. In his final will, Hance named Hooker, Bentham, and Daniel Oliver, Keeper of Kew Herbarium, to be the representatives of his interests in the sale of his herbarium after his death. On the efforts to obtain promotion for Hance, see Kew: Misc. Reports. 4.4. China. Hance. 1882–88, ff. 1–5, 23–25, 49.

52. Kew: Directors Correspondence. Chinese Letters, 57 (73)–(77).

53. Kew: Misc. Reports. 4.4. China. Economic Products. I. ff. 229–230.

54. Kew: Chinese and Japanese Letters, 150 (574–575). See also Ford's letters to Kew, 150 (217–218), 150 (221).

55. Kew: Chinese and Japanese Letters, 150 (574–575).

56. Jim Endersby, "Expertise: Joseph Hooker's Australasian Correspondence with William Colenso and Ronald Gunn," *Pacific Science* 55 (2001): 343–358.

57. There was no serious debate over Darwinism among British naturalists in China or between them and their colleagues in Britain. In contrast, the response of French naturalists in China, most of whom were Catholic missionaries, to Darwinism was very different. Like the scientific community in France, they ignored it at first. In the last decades of the nineteenth century, their attitude turned hostile. An authority on Chinese molluscs, the Jesuit missionary Pierre Marie Heude (1836–1902) at Xujiahui (Zi-ka-wei or Sicawei) attacked Darwinism and adopted an idiosyncratic theory of morphology and systematics in his work. More on Heude, see Chapter 4.

58. Frederick Burkhardt et al., eds., *The Correspondence of Charles Darwin*, vol. 8 (Cambridge, England: Cambridge University Press, 1985), 556.

59. H. F. Hance, "Remarks on the Modern Tendency to Combine Species," *The Journal of Botany, British and Foreign* (also known as *Seemann's Journal of Botany*) 4 (1866): 84–86.

60. Kew. Directors Correspondence. 150 (504–505), (508–511). These were friendly differences. Hance later described himself as "having long since felt compelled to accept the doctrine of Evolution as the master key to the varied affinities of existing organisms," and said that he "fully believe[d] in their common descent." H. F. Hance, "On Silk-worm Oaks," *The China Review* 6 (1877): 207–208.

61. The consular officers and the consular establishments in China are conveniently listed in Coates, *The China Consuls*, 489–490.

62. Wright, *Hart and the Chinese Customs*, and the explanatory essays in Katherine F. Brunner, John K. Fairbank, and Richard J. Smith, eds., *Entering China's Service: Robert Hart's Journals, 1854–1863* (Cambridge, Mass.: Harvard University Press, 1986) and idem, *Robert Hart and China's Early Modernization: Robert Hart's Journals, 1863–1866* (Cambridge, Mass.: Harvard University Press, 1991).

63. Wu Chouyi, *Qingmo Shanghai zujie shehui* (Taipei: Wenshizhe chubanshe, 1978), 55.

64. Coates, *The China Consuls*, 76–80; Wright, *Hart and the Chinese Customs*, 260–271. Robert Hart frequently recorded his thoughts about recruiting new members in his journals and letters. See Hart's journals in note 62 and J. K. Fairbank, K. F. Bruner, and E. M. Matheson, eds., *The I. G. in Peking: Letters of Rob-*

ert Hart, *Chinese Maritime Customs, 1868–1907*, 2 vols. (Cambridge, Mass.: Harvard University Press, 1975).

65. Shang-Jen Li, "Natural History of Parasitic Disease: Patrick Manson's Philosophical Method," *Isis* 93 (2002): 206–228; Douglas M. Haynes, *Imperial Medicine: Patrick Manson and the Conquest of Tropical Disease, 1844–1929* (Philadelphia: University of Pennsylvania Press, 2001).

66. Robert Fynn, *British Consuls Abroad; their Origin, Rank and Privileges, Duties, Jurisdiction and Emoluments; Including the Laws, Orders in Council, and Instructions by Which They are Governed, As Well As Those Relating to Ship-Owners and Merchants in Their Connection with Consuls* (London: E. Wilson, 1846), 30–31, describes the basic personnel structure of the China consulates, which would greatly expand in the rest of the century.

67. As a contemporary witness, Bretschneider knew well this fact. See his *History*, 631.

68. One of the criteria for the entrants was "a public-school and something beyond." See Coates, *The China Consuls*, 78.

69. As a student interpreter, Swinhoe set aside the standard textbooks and tackled a Chinese natural history book for his language lessons. See Coates, *The China Consuls*, 98. Newly arrived in Hong Kong, Alabaster already began gathering "various grasses" and had "some thoughts of collecting and writing a book about them myself," though he would later become a dedicated diplomat and leave little time for botany. In SOAS. Chaloner Alabaster Papers. Diary 1. MS 380451/1, 18, 19, 28 October 1855, etc. G. M. H. Playfair, who was from a prominent medical-colonial family, already showed an interest in botany in his first letters to Kew, written shortly after his arrival in China. Playfair to Hooker, Kew: Chinese and Japanese Letters, 151 (875–879). For Bowra's early interest in botany, see, for example, July 18 in his diary for 1863, SOAS. Bowra Papers. MS English. 201813. Box 2. No. 7.

70. Robert Swinhoe, "Ornithological Ramble in Foochow, in December 1861," *Ibis* 4 (1862): 257.

71. Kew: Augustine Henry. Letters to H. B. Morse, 1893–1909, ff. 3–5; 7–9; 10–12; 17–18; 19; 29; 34–35. See also John King Fairbank, Martha Henderson Coolidge, and Richard J. Smith, *H. B. Morse: Customs Commissioner and Historian of China* (Lexington: University of Kentucky Press, 1995), 117–118.

72. These were daily routines of Augustine Henry and his colleagues in Yichang. See his diaries in NBG. Augustine Henry Papers. 582.095. Henry's other motivation for studying botany was his growing interest in Chinese *materia medica*. See Chapter 4.

73. Hancock to Hooker, 22 April 1884, Kew: Chinese and Japanese Letters, 151 (470–472). See also 151 (475), (476), (481). On Hooker's scientific travels, see Ray Desmond, *Sir Joseph Dalton Hooker: Traveller and Plant Collector* (Woodbridge, England: Antique Collectors' Club, 1999). For science and mountaineering, see Bruce Hevly, "The Heroic Science of Glacier Motion," in *Osiris* 11 (1996): 66–86. Peter H. Hansen discusses mountaineering and masculine imperialism in "Albert Smith, the Alpine Club, and the Invention of Mountaineering in Mid-Victorian Britain," *Journal of British Studies* 34 (1995): 300–324; idem, "British Mountaineering, 1850–1914" (Ph.D. diss., Harvard University, 1991).

74. Hancock's intense interest in natural history can be seen in his "Notes on the Physical Geography, Flora, Fauna, etc., of Northern Formosa, with Compari-

sons between that District and Hainan and Other Parts of China," which was attached to his *Tamsui Trade Report for the Year 1881* (Tamsui: Customs House, 1882). Ibid., 31–38, describes his encounter with the aborigines. Robert Swinhoe, *Notes on the Ethnology of Formosa* (London: Frederick Bell, 1863), originally read before the British Association for the Advancement of Science, August 1863. The issue of British imperialism and masculinity is discussed, in a somewhat speculative mode, in Richard Phillips, *Mapping Men and Empire: A Geography of Adventure* (London: Routledge, 1997), esp. chs. 3 and 4; G. Dawson, *Soldier Heroes: British Adventure, Empire and the Imagining of Masculinities* (London: Routledge, 1994); Jonathan Rutherford, *Forever England: Reflections on Race, Masculinity and Empire* (London: Lawrence & Wishart, 1997). See also the works by Hevly and Hansen in note 73.

75. Hancock to Hooker, 22 April 1884, Kew: Chinese and Japanese Letters, 151 (471). See also "Shanghai Museum, Report of the Curator for the Year 1880," *JNCB*, n.s., 15 (1880): xxiv; Bretschneider, *History*, 631. In addition to the annual reports from the various ports, the Chinese Maritime Customs also published special series concerning scientific subjects. For example, *List of Chinese Medicines*. China. Imperial Maritime Customs, III. Miscellaneous Series, No. 17 (Shanghai: Statistical Department of the Inspector General of Customs, 1889); *An Epitome of the Reports of the Medical Officers to the Chinese Imperial Maritime Customs Service from 1871 to 1882: With Chapters on the History of Medicine in China; Materia Medica; Epidemics; Famines; Ethnology; and Chronology in Relation to Medicine and Public Health* (London: Baillère, Tindall and Cox, 1884); *Silk*, China. Imperial Maritime Customs, II. Special Series, No. 3 (Shanghai: Department of the Inspectorate General, 1881); *Special Catalogue of the Chinese Collection of Exhibits for the International Fisheries Exhibition, London, 1883* (Shanghai: Statistical Department of the Inspector General, 1883).

76. Kew: Chinese and Japanese Letters, 151 (596–97), (621), (627), (629).

77. Hart was quite conscientious about his role as an employee of the Chinese government and required his subordinates to be the same, though conflicting feelings and interests were sometimes inevitable. For the details, see Hart's journals and letters cited in note 62 and Wright, *Hart and the Chinese Customs*, 261–262.

78. Henry to C. S. Sargent, 10 May 1899, NBG. Augustine Henry Papers, in the file "C. S. Sargent."

79. La Touche to Oldfield Thomas, 28 July 1898, NHML: Curator of Mammals. Correspondence. DF232/6, f. 512. G. A. Boulenger, "On a Collection of Reptiles and Batrachians Made by Mr. J. D. La Touche in N. W. Fokien, China," *Proc. Zool. Soc.* (1899): 161.

80. Copies of Kew's applications to the Foreign Office for assistance and replies from the China consuls are collected in Kew: Misc. Reports. 4.4. China: Foods, Medicines, & Woods, 1869–1914; Misc. Reports. 4.4. China: Economic Products, I; Misc. Reports. 4.4. China. Economic Products, II; Misc. Reports 4.4: China & Tibet, Misc. 1861–1924; Misc. Reports. 4.4. China. Economic Products. Insect White Wax; Kew: Misc. Reports. China. Plant Collections. Cultural Products, etc. 1853–1914.

81. Philip B. Hall, "Robert Swinhoe (1836–1877), FRS, FZS, FRGS: A Victorian Naturalist in Treaty Port China," *The Geographical Journal* 153 (March 1987): 37–47. For example, Thomas Watters, a consular officer and renowned Buddhism scholar, went through fourteen offices in his thirty-year career. See *The Foreign*

Office List, Containing Diplomatic and Consular Appointments, &c. (London, 1896), 223.

82. Not surprisingly, some transfers could work against natural historical research. William Hancock was posted to Hankou in 1890 and he complained that "[during] fifteen years residence in China" he has never resided in "a more uninteresting & (botanically) unproductive region than Hankow." Hancock to Hooker, 27 April 1890, Kew: Chinese and Japanese Letters, 151 (481). But at a time when so much of the flora and fauna of China remained unknown, new moves usually meant good opportunities.

83. Oldham to William Hooker, 19 March 1864, Kew: Kew Collectors. Oldham, 1861–1864, IV, ff. 33–36.

84. For Swinhoe's scientific research in Taiwan, see Zhang Yuteng, "Yingguo bowuxuejia Shiwenhou zai Taiwan de zilanshi diaocha jingguo ji xiangguan shiliao," *Taiwan shi yanjiu* 1 (1993): 132–151.

85. The letters are in the file NBG. Augustine Henry. "Letters to Henry." See also the entries under the dates in his diary, NBG. Augustine Henry. 582.095.

86. About Swinhoe's paying more attention to natural history than to his wife, see Smith et al., eds., *Robert Hart and China's Early Modernization*, 256.

87. Robert Swinhoe, "On the White Stork of Japan," *Proc. Zool. Soc.* (1873): 512–514.

88. Robert Swinhoe, "On Chinese Deer, with the Description of an Apparently New Species," *Proc. Zool. Soc.* (1873): 574. Kopsch was also similarly rewarded for sending Swinhoe bird specimens. For example, Swinhoe, "On a New Species of Nettapus (Cotton-Teal) from the River Yangtze, China," *Annals and Magazines of Natural History*, sr. 4., 11 (1873): 15–17.

89. Robert Swinhoe to A. Günther, 23 April 1866, NHML: Z. Keeper's Archives. 1.8. Letters. 1858–75. SM-Z, No. 284. Swinhoe forwarded the fish to the British Museum (Natural History) for identification.

90. William Cooper to Thiselton-Dyer, 2 November 1885, Kew: Chinese and Japanese Letters, 151 (138), (142–43).

91. George Busk, "Notes on the Cranial and Dental Characters of the Northern and Southern Tigers and Leopards of China as Affording Marks of Their Specific Distribution," *Proc. Zool. Soc.* (1874): 146–150.

92. Robert Swinhoe, "On the Mammals of the Island of Formosa," *Proc. Zool. Soc.* (1862): 347–365, 351.

93. Swinhoe, "On the Mammals of the Island of Formosa," 362. According to Swinhoe, the mandarin "had [the goat blood] spread . . . in small cakes, dried, and powdered, and then stowed it carefully away in his medicine-chest."

94. Henry Hance, "On the Source of the China Root of Commerce," *The Journal of Botany, British and Foreign* (otherwise known as *Seemann's Journal of Botany*) 10 (1872): 102–103.

95. Bretschneider, *Notes on Some Botanical Questions Connected with the Export Trade of China* (Peking, 1880).

96. Bretschneider, "Botanicon Sinicum: Notes on Chinese Botany from Native and Western Sources, Part III," *JNCB* 29 (1895): 11–12; idem, *On Chinese Silkworm Trees* (Peking: Privately printed, 1881), 1–2. See also his *Notes on Some Botanical Questions*, 1.

97. Henry Fletcher Hance, "On the Source of the *Radix Galangae Minoris* of

Pharmacologists," *The American Journal of Pharmacy* (September 1871): 404–410; Daniel Hanbury, "Historical Notes on the Radix Galangae of Pharmacy," *American Journal of Pharmacy* (October 1871): 452–456.

98. Biographic information about these and many other French missionaries can be found in Bretschneider, *History*, 824–929.

99. The Xujiahui (also known as Zikawei or Sicawei) station also had an excellent observatory. See Lewis Pyenson, *Civilizing Mission: Exact Science and French Overseas Expansion, 1830–1940* (Baltimore: Johns Hopkins University Press, 1993), 155–206.

100. Medhurst, *The Foreigner in Far Cathay*, 32–35; Armand David, a French missionary-naturalist, was scornful of the Protestant approach: "These men, generally encumbered with a family, can scarcely undertake hazardous enterprises." Armand David, *Abbé David's Diary* (Cambridge, Mass.: Harvard University Press, 1949), 193.

101. On the China Inland Mission, see Leslie T. Lyall, *A Passion for the Impossible: The China Inland Mission, 1865–1965* (London: Hodder and Stoughton, 1965) and A. J. Broomhall, *Hudson Taylor and China's Open Century*, 7 vols. (London: The Overseas Missionary Fellowship, 1981–1989).

102. Father Hugh to Oldfield Thomas, 18 February 1899. NHML: Curator of Mammals. Correspondence. DF232/5, No. 234–235.

103. Of them Williams, McGowan, and Henry were American; Faber, German. The other ones were British. Basic biographic information about them can be found in Bretschneider, *History*.

104. John Ross collected in southern Manchuria. See his letters to Hooker, Kew: Chinese and Japanese Letters, 151 (916)–(918) and J. G. Baker, "A Contribution to the Flora of Northern China," *Journal of the Linnean Society* 42 (1880): 375–390. On Alexander Williamson, see his *Journeys in North China, Manchuria, and Eastern Mongolia*, vol. 2 (London: Smith, Elder & Co, 1870), "Appendix D. List of Plants from Shan-tung, collected by the Rev. A. Williamson." B. C. Henry, an American missionary, collected for Henry Hance on his missionary itineraries. Henry, *Ling-Nam or Interior Views of Southern China*, 123–124; Kew: Chinese and Japanese Letters, 150 (338–339). E. Faber, a scholarly German missionary, collected for Hance and Charles Ford. See Bretschneider, *History*, 954–959; Kew: Chinese and Japanese Letters, 150 (353–354), (572). F. Porter Smith, who was a British missionary doctor in Hankou, compiled *Contributions towards the Materia Medica and Natural History of China: For the Use of Medical Missionaries and Native Medical Students* (Shanghai: American Presbyterian Press, 1871).

105. There was also a population of working-class Westerners in Shanghai (and, for that matter, Hong Kong), but whether they collected specimens or studied China's natural history at all is no longer traceable. Anne Secord has shown that at least some artisans in Britain took up natural history with enthusiasm. Anne Secord, "Artisan Botany," in N. Jardine et al., eds., *Cultures of Natural History*, 378–393; idem, "Science in the Pub: Artisan Botanists in Early Nineteenth-Century Lancashire," *History of Science* 32 (1994): 269–315; idem, "Corresponding Interests: Artisans and Gentlemen in Natural History Exchange Networks," *British Journal for the History of Science* 27 (1994): 383–408.

106. James Troyer, "John Charles Bowring (1821–1893): Contributions of a Merchant to Natural History," *Archives of Natural History* 10 (1982): 515–529. See

also [Anon.], *The History of the Collections Contained in the Natural History Departments of the British Museum*, vol. 2 (London: The Trustees of the British Museum, 1904–12), 581, which states, "In 1862 Bowring presented to the [British] Museum his entire collection of Coleoptera, consisting of about 230,000 specimens. Bowring himself collected energetically while abroad in China, &c., and also employed persons to collect for him." Large proportions of his collections had been purchased from entomologists in Europe.

107. George Lanning, "Thos. W. Kingsmill," *JNCB* 41 (1910): 116–118; Robert Swinhoe, "Zoological Notes of a Journey from Canton to Peking and Kalgan," *Proc. Zool. Soc.* (1870): 427–451 (on pp. 428–429).

108. Bretschneider, *History*, 720–723.

109. Ferdinand von Richthofen, *Letters to the Shanghai General Chamber of Commerce* (Shanghai, 1875). Wang Gen-yuan and Michel Gert, "The German Scholar Ferdinand von Richthofen and Geology in China," in Wang Hungchen et al., eds., *Interchange of Geoscience Ideas Between the East and the West (Proceedings of the XVth International Symposium of INHIGEO)* (Wuhan: China University of Geoscience Press, 1991), 47–54.

110. See, for example, A. A. Fauvel, "Shanghai Museum, Report of the Curator for the Year 1878," *JNCB*, n.s., 13 (1878): xvii–xviii; D. C. Jansen, "Shanghai Museum, Report of the Curator for the Year 1880," *JNCB*, n.s., 15 (1880): xxiii–xxiv.

111. H. B. Morse, "Report of the Council on the Proposed Trade and Commerce Museum," *JNCB*, n.s., 23 (1888): 49–53.

112. For example, Henry Hance, "On Silk-Worm Oaks," *China Review* 6 (1877): 207–208; idem, "On the Sources of the 'China Matting' of Commerce," *Journal of Botany*, n.s., 8 (1879): 99–105; idem, "On a New Chinese Caryota," *Journal of Botany*, n.s., 8 (1879): 174–177.

113. P. L. Sclater, *Guide to the Gardens of the Zoological Society of London* (London: Bradbury, Evans, and Co., 1872), 39; A. D. Bartlett, "Description of Chinese Sheep Sent to H. R. H. Prince Albert by Rutherford Alcock," *Proc. Zool. Soc.* (1857): 104–107. Overall, however, the British made no systematic attempts to acclimatize Chinese animals, and there is no evidence that there was any notable animal trade between China and Britain (or British colonies). On the animal trade between India and Britain, see Christine Brandon-Jones, "Edward Blyth, Charles Darwin, and the Animal Trade in Nineteenth-Century India and Britain," *Journal of the History of Biology* 30 (1997): 145–178. Generally speaking, the French were more active than the British in acclimatizing animals. See, for example, Michael Osborne, "The Société Zoologique d'Acclimatation and the New French Empire: Science and Political Economy," in P. Petitjean et al., *Science and Empires* (Amsterdam: Kluwer Academic Publishers, 1992), 299–306; idem, *Nature, the Exotic, and the Science of French Colonialism* (Bloomington: Indiana University Press, 1994). The French in China, for example, tried to introduce Chinese fish into France for acclimatization. See, for example, P. Dabry de Thiersant, *La pisciculture et la pêche en Chine* (Paris: Librairie de G. Massion, 1872). Dabry de Thiersant was French consul at Shanghai, honorary member of the Société d'Acclimatation, and an active collector of zoological specimens.

114. Kew: Misc. Reports. 4.4. China. Economic Products. II., ff. 284–285, 286–288, 289–295.

115. Robert Fortune's endeavors for tea hunting were described in his *Journey to*

the *Tea Countries of China* (London: Midmay Books, 1987 [1852]) and *A Residence among the Chinese: Inland, on the Coast, and at Sea* (London: John Murray, 1857). The East India Company had previously sent George James Gordon to China to collect tea seeds and recruit Chinese tea manufacturers to India. With the help of the German missionary Karl Gutzlaff and others, he accomplished his mission with some success. See *Tea Cultivation (India): Copy of Papers received from India relating to the Measures adopted for introducing the Cultivation of the Tea Plant with the British Possessions in India*, comp. James Melveill (Copy by the House of Commons, 1839), 5–6, 29–46, 57–63, 72–74, 79–80, 87–90, 117–118; Percival Griffiths, *The History of the Indian Tea Industry* (London: Weidenfeld and Nicolson, 1967), 34–58, 64–69, 78–81, 90. The Indian industry went through some bad times in the third quarter of the nineteenth century, but the reason was not technological.

116. Kew: Chinese and Japanese Letters, 150 (285).

117. *Gardeners Chronicle*, 22 June 1844, 405.

118. Robert Fortune, *Three Years' Wanderings in the Northern Provinces of China* (London: John Murray, 1847), 55.

119. Fortune, *A Journey to the Tea Countries*, 130–132. Fortune either recounted the same incidents twice or watched Aching pack seeds again in this second journey.

120. Public Records Office (Hong Kong). CO 129/202, 129/206.

121. B. C. Henry, *Ling-Nam or Interior View of Southern China*, 102. This incident was hardly unique. "In reference to the Chinese jealousy of parting with their seeds," Ford told Thiselton-Dyer some years later, "they will not allow seeds even of the fan palm . . . to pass the Customs stations in the district where the palms are cultivated." Kew: Chinese and Japanese Letters, 150 (417–418).

122. Fortune, *A Residence among the Chinese*, 109–115.

123. I agree with John M. MacKenzie and others that orientalism as a historical phenomenon was rich in connotation and that European representations of the Orient were not as uniform and static as portrayed by Edward Said. John M. MacKenzie, *Orientalism: History, Theory and the Arts* (Manchester: University of Manchester Press, 1995); Edward Said, *Orientalism* (New York: Pantheon, 1978). It is probably true, too, that China occupied a somewhat special place in Western representations of other civilizations. Jonathan Spence, "Western Perceptions of China from the Late Sixteenth Century to the Present," in *Heritage of China: Contemporary Perspectives on Chinese Civilization*, ed. Paul S. Ropp (Berkeley: University of California Press, 1990), 1–14, and his *The Chan's Continent: China in Western Minds* (New York: Norton, 1998). Throughout the book, I use the term "orientalism" primarily to denote a principal way of representing non-Western peoples and civilizations, one which depended heavily on the language of "othering" and which prevailed in certain kinds of narrative.

124. The concept of resistance is of course not limited to the context of colonialism/imperialism and has been used widely to interpret power practice in everyday life. See, for example, John Fiske, *Understanding Popular Culture* (Boston: Unwin Hyman, 1989). But for the present study, the most relevant context is probably a colonial/imperial one. My thinking has involved, among others, the following works. Gyan Prakash, ed., *After Colonialism: Imperial Histories and Postcolonial Displacements* (Princeton: Princeton University Press, 1995); Edward Said, *Culture and Imperialism* (New York: Alfred A. Knopf, 1993), ch. 3; Homi K. Bhabha, *The*

Location of Culture (London: Routledge, 1994). See also "Introduction" to Nicholas Dirks, Geoff Eley, and Sherry Ortner, eds., *Culture/Power/History: A Reader in Contemporary Social Theory* (Princeton, N. J.: Princeton University Press, 1994), 3–45; Sherry Ortner, "Resistance and the Problem of Ethnographic Refusal," in Terence J. McDonald, ed., *The Historical Turn in the Human Sciences* (Ann Arbor: University of Michigan Press, 1996), 304.

125. I borrowed the phrase from James C. Scott, *Weapons of the Weak: Everyday Forms of Peasant Resistance* (New Haven: Yale University Press, 1985). See also his wide-ranging *Domination and the Arts of Resistance: Hidden Transcripts* (New Haven: Yale University Press, 1990). It should be noted that Scott is primarily concerned with power relations between social classes.

126. I have been influenced by Sherry B. Ortner, "Thick Resistance: Death and the Cultural Construction of Agency in Himalayan Mountaineering," *Representations*, no. 59 (Summer 1997): 135–162; Lydia Liu, *Translingual Practice: Literature, National Culture, and Translated Modernity—China, 1900–1937* (Stanford: Stanford University Press, 1995).

127. Fortune, *Three Years' Wanderings in China*, 3.

128. Fortune, *A Residence among the Chinese*, 206.

129. For example, Shapin, *A Social History of Truth*; Lorraine Daston, "The Moral Economy of Science," in *Constructing Knowledge in the History of Science*, ed. Arnold Thackray; *Osiris* 10 (1995): 3–24; idem, "Baconian Facts, Academic Civility, and the Prehistory of Objectivity," *Annals of Scholarship* 8, nos. 3–4 (1991): 337–364; Mary Poovey, *A History of the Modern Fact: Problems of Knowledge in the Sciences of Wealth and Society* (Chicago: University of Chicago Press, 1998).

130. See note 75.

131. Thomas Taylor Meadows, "Report on the Consular District of New-Chuang, with the Particular Reference to Its Commercial Capabilities," 1. The report is collected in *Commercial Reports from Her Majesty's Consuls in China for the Year 1862* (Printed to both Houses of Parliament, 1864), 1–21. Victor Hilts, *Statist and Statistician: Three Studies in the History of Nineteenth-Century English Statistical Thought* (New York: Arno Press, 1981); idem, "Aliis Exterendum, or the Origins of the Statistical Society of London," *Isis* 69 (1978): 21–43; Theodore M. Porter, *The Rise of Statistical Thinking, 1820–1900* (Princeton: Princeton University Press, 1986); Poovey, *A History of the Modern Fact*, chs. 6 and 7.

132. S. Wells Williams, *The Middle Kingdom; a Survey of the Geography, Government, Literature, Social Life, Arts, and History of the Chinese Empire and Its Inhabitants*, vol. 2 (New York: Charles Scribner's Sons, 1913 [1882]), 65; John Davis, *China: A General Description of that Empire and Its Inhabitants*, rev. ed., 2 vols. (London: John Murray, 1857), vol. 2, 221. Michael Adas has usefully reminded us that science and technology were among the most important criteria with which nineteenth-century Westerners ranked other peoples, though he unnecessarily downplays the tendency of the Westerners to interpret the hierarchy of peoples in terms of their physical and mental attributes. Michael Adas, *Machines as the Measure of Men: Science, Technology, and Ideologies of Western Dominance* (Ithaca: Cornell University Press, 1989), 338–342.

133. I have benefited from the insights in Pratt, *The Imperial Eye* and Stephen Greenblatt's *Marvelous Possessions: The Wonder of the New World* (Chicago: University of Chicago Press, 1991).

134. Identifying this vision alone, of course, does not fully explain the methods and strategies the naturalists adopted to interpret the information collected from the Chinese. For more on this issue, see Chapter 4.

4. Sinology and Natural History

1. The professionalization of science in nineteenth-century Britain is a longstanding historiographic issue. Two recent interventions in the historiography are particularly pertinent here: Richard Drayton, *Nature's Government* (see Chapter 3, note 5) and a special issue of the *Journal of the History of Biology* 34 (2001), which provides a more multifaceted view than the former.

2. The essays in Ming Wilson and John Cayley, eds., *Europe Studies China* (London: Han-Shan Tang Books, 1995) together provide a panoramic survey of the history of sinology in Europe. See also Herbert Franke, *Sinology in German Universities* (Wiesbaden: Franz Steiner, 1968), 1–11; T. H. Barrett, *Singular Listlessness: A Short History of Chinese Books and British Scholars* (London: Wellsweep Press, 1989), 19–75; J. J. L. Duyvendak, *Holland's Contribution to Chinese Studies* (London: The China Society, 1950), 3–23; Ch'en Yao-sheng and Paul S. Y. Hsiao, *Sinology in the United Kingdom and Germany* (Honolulu: East-West Center, 1967); Norman J. Girardot, *The Victorian Translation of China: James Legge's Oriental Pilgrimage* (Berkeley: University of California Press, 2002), esp. 1–16.

3. Edward Said, *Orientalism* (see Chapter 3, note 123); Raymond Schwab, *The Oriental Renaissance: Europe's Rediscovery of India and the East, 1680–1880* (New York: Columbia University Press, 1984).

4. The particular issue has been discussed in Chapter 3.

5. Many of the essays in Jardine, Secord, and Spray, eds., *Cultures of Natural History* (see "Introduction," note 4) and in Bernard Lightman, ed., *Victorian Science in Context* (see Chapter 2, note 4) concern the practice of natural history. For the literature on fieldwork, see Chapter 5. On the natural history museum, see Mary Winsor, *Reading the Shape of Nature: Comparative Zoology at the Agassiz Museum* (Chicago: University of Chicago Press, 1991); Sally G. Kohlstedt, "Museums: Revisiting Sites in the History of the Natural Sciences," *Journal of the History of Biology* 28 (1995): 151–166; Sophie Forgan, "The Architecture of Display: Museums, Universities and Objects in Nineteenth-Century Britain," *History of Science* 23 (1994): 139–162; Timothy Lenoir and Cheryl Lynn Ross, "The Naturalized History Museum," in Peter Galison and David Stump, eds., *The Disunity of Science: Boundaries, Contexts, and Power* (Stanford: Stanford University Press, 1996), 370–397.

6. William Ashworth's "Emblematic Natural History of the Renaissance," in Jardine, Secord, and Spray, eds., *Cultures of Natural History*, 17–37, takes the mid-seventeenth-century as the turning point. On the other hand, some see Buffon and Linnaeus as pivotal figures. See Rhonda Rappaport, *When Geologists Were Historians, 1665–1750* (Ithaca: Cornell University Press, 1997), ch. 3; John Lyon and Phillip R. Sloan, *From Natural History to the History of Nature: Readings from Buffon and His Critics* (Notre Dame: University of Notre Dame Press, 1981), ch. 1; Sten Lindroth, "Two Faces of Linnaeus," in Tore Frängsmyr, ed., *Linnaeus: The Man and His Work* (Canton, Mass.: Science History Publications, 1994), 1–62.

7. Several scholars have pointed out the connections between humanism and natural history during an earlier period. For example, Barbara Shapiro, "History

and Natural History in Sixteenth- and Seventeenth-Century England: An Essay on the Relationship between Humanism and Science," in Barbara Shapiro and Robert Frank, eds., *English Scientific Virtuosi in the 16th and 17th Centuries* (Los Angeles: Clark Library, 1979), 3–55; Paula Findlen, *Possessing Nature: Museums, Collecting, and Scientific Culture in Early Modern Italy* (Berkeley: University of California Press, 1994); William Ashworth suggests the importance of antiquarianism in natural history; see his "Natural History and the Emblematic World View," in David Lindberg and Robert Westman, eds., *Reappraisals of the Scientific Revolution* (Chicago: University of Chicago Press, 1990), 303–332. Joseph Levine, *Dr. Woodward's Shield: History, Science, and Satire in Augustan England* (Ithaca: Cornell University Press, 1977); idem, "Natural History and the New Philosophy: Bacon, Harvey, and the Two Cultures," in *Humanism and History* (Ithaca: Cornell University Press, 1987), 123–154; Paolo Rossi, *The Dark Abyss of Time: The History of the Earth and the History of Nations from Hooke to Vico* (Chicago: University of Chicago Press, 1984), parts II and III; Ann Blair, "Humanist Methods in Natural Philosophy: The Common Place Book," *Journal of the History of Ideas* 53 (1992): 541–551; Jerome Bylebyl, "The School of Padua: Humanistic Medicine in the Sixteenth Century," in Charles Webster, ed., *Health, Medicine and Morality in the Sixteenth Century* (Cambridge, England: Cambridge University Press, 1979), 335–370. Anthony Grafton has emphasized the broad and lasting influence of humanist scholarship in a different context. See, for example, his *Defenders of the Text: The Traditions of Scholarship in an Age of Science, 1450–1800* (Cambridge, Mass.: Harvard University Press, 1991).

8. For example, Jacques Gernet, *China and the Christian Impact* (Cambridge, England: Cambridge University Press, 1987), esp. 238–247; Alain Peyrefitte, *The Immobile Empire* (cited in Chapter 1, note 42).

9. We have seen in the last chapter how orientalism and the modern conception of factual knowledge converged in the Westerners' definition of the Other.

10. Benjamin Schwartz, "Culture, Modernity, and Nationalism—Further Reflections," *Daedalus* 122 (Summer 1993): 207–226, esp. 207–208. Schwartz criticizes and modifies this view.

11. I have benefited from the insights in a number of otherwise very different works. Lydia Liu, *Translingual Practice* (see Chapter 3, note 126), 1–44; James Hevia, *Cherishing Men from Afar* (see Chapter 1, note 42); G. E. R. Lloyd, *Demystifying Mentalities* (New York: Cambridge University Press, 1990). The controversy around Hevia's book concerns to some extent this methodological issue. It seems to me that one does not have to embrace Hevia's interpretation of the Macartney Embassy to appreciate some of his methodological concerns. Joseph Esherick, "Cherishing Sources from Afar," in *Modern China* 24 (1998): 135–161; James Hevia, "Postpolemical Historiography," *Modern China* 24 (1998): 319–327; Joseph Esherick, "Tradutore, Traditore," ibid., 328–332.

12. By comparison, the Chinese intellectuals' roles in translating Western science into late Qing China are much better documented and studied. The best survey is Xiong Yuezhi, *Xixue dongjian yu wan Qing shehui* (Shanghai: Shanghai renmin chubanshe, 1994). See also David Wright, *Translating Science: The Transmission of Western Chemistry into Late Imperial China, 1840–1900* (Leiden: Brill, 2000); John Rear-Anderson, *The Study of Change: Chemistry in China, 1840–1949* (New York: Cambridge University Press, 1991), chs. 1 and 2; Paul Cohen, *Between Tradition and Modernity: Wang Tao and Reform in Late Ching China* (Cambridge, Mass.: Har-

vard University Press, 1974); James Pusey, *China and Charles Darwin* (Cambridge, Mass.: Harvard University Press, 1983); Wang Yangzong and Fan Hongye, *Xixue dongjian—kexue zai Zhongguo de chuanbo* (Changsha: Hunan kexuejishu chubanshe, 2000); Sakade Yoshinobu, *Chūgoku kindai no shisō to kagaku* (Kyoto: Dōhōsha, 1983), ch. 4; Bridie Andrews, "Tuberculosis and the Assimilation of Germ Theory in China, 1895-1937," *Journal of the History of Medicine and Allied Sciences* 52 (January 1997): 114-157.

13. Emil Bretschneider, "Botanicon Sinicum: Notes on Chinese Botany from Native and Western Sources," *JNCB* 16 (1882): 19.

14. Zhang Hailin, *Wang Tao pingzhuan* (Nanjing: Nanjing daxue chubanshe, 1993), 100-105. Even Girardot's biography of James Legge, which exalts Legge's place in the history of European sinology, admits the contributions of Legge's Chinese collaborators. Girardot, *The Victorian Translation of China*, 356-357.

15. The literature on Jesuits in China is legion. See, for example, Willard Peterson, "Learning from the Heaven: The Introduction of Christianity and Other Western Ideas into Late Ming China," in *The Cambridge History of China*, vol. 8, part II (Cambridge, England: Cambridge University Press, 1998), 789-839; Gernet, *China and the Christian Impact*; Charles Ronan and Bonnie Oh, *East Meets West: The Jesuits in China* (Chicago: Loyola University Press, 1988); Jonathan Spence, *The Memory Palace of Matteo Ricci* (London: Faber & Faber, 1985). John Young, *Confucianism and Christianity, the First Encounter* (Hong Kong: Hong Kong University Press, 1983); Han Qi, *Zhongguo kexue jishu xichuan ji qi yingxiang* (Shijiazhuang: Hebei renmin chubanshe, 1999).

16. Matthew Ricci, *China in the 16th Century: The Journals of Matthew Ricci, 1583-1610* (New York: Random House, 1953), 28.

17. The Chinese interpreters were Catholic converts brought to Europe by the Jesuits. See Barrett, *Singular Listlessness*, 37-38, 49; Jonathan Spence, "The Paris Years of Arcadio Huang," in his *China Roundabout* (New York: W. W. Norton, 1992), 11-24; idem, *The Question of Hu* (New York: Vintage Books, 1989).

18. For example, Knud Lundbaek, *T. S. Bayer (1694-1738), Pioneer Sinologist* (London: Curzon Press, 1986), 31-140.

19. John Bold, "John Webb: Composite Capitals and the Chinese Language," *Oxford Art Journal* 4 (1981): 9-17. Rossi, *The Dark Abyss of Time*, 137-144. David Mungello, *Curious Land: Jesuit Accommodation and the Origins of Sinology* (Stuttgart: Franz Steiner Verlag Wiesbaden GMBH, 1985), chs. 4-6.

20. Antonello Gerbi, *The Dispute of the New World: The History of a Polemic, 1750-1900* (Pittsburgh: University of Pittsburgh Press, 1973), 150-154.

21. Mary Slaughter, *Universal Languages and Scientific Taxonomy in the Seventeenth Century* (Cambridge, England: Cambridge University Press, 1982), 112-113; James Knowlson, *Universal Language Schemes in England and France, 1600-1800* (Toronto: University of Toronto Press, 1975), 25-27; David Mungello, *Leibniz and Confucianism: The Search for Accord* (Honolulu: University of Hawaii Press, 1977), 43-65; idem, *Curious Land: Jesuit Accommodation and the Origins of Sinology*, ch. 6.

22. C. A. Wells, *The Origin of Language: Aspects of the Discussion from Condillac to Wundt* (Le Salle, Ill.: Open Court, 1987), 61-62.

23. L. C. Goodrich, "Boym and Boymiae," *T'oung Pao* 57 (1971): 135; J. Roi, "Les missionnaires de Chine et la botanique," *Collectanea Commissionis Synodalis in*

Sinis 11 (1938): 695–706; Edward Kajdnski, "Receptarum Sinensium Liber of Michael Boym," *Janus* 73 (1990): 105–124; Paul Pelliot, "Michael Boym," *T'oung Pao* 30 (1933): 95–151. Robert Chabrie, *Michel Boym. Jésuite polonais et la fin des Ming en Chine* (Paris: Bossuet, 1933) remains the fullest account of his life. Pan Jixing, *Zhongwai kexue zhi jiaoliu* (Hong Kong: Chinese University Press, 1993), 479.

24. For the musk deer, see Timothy James Billings, "Illustrating China: Emblematic Autopsy and the Catachresis of Cathay" (Ph.D. diss., Cornell University, 1997), 243–265; Martha Baldwin, "The Snakestone Experiments: An Early Modern Medical Debate," *Isis* 86 (1995): 394–418.

25. Clifford Foust, *Rhubarb: The Wondrous Drug* (Princeton: Princeton University Press, 1992); Daniel Carey, "Compiling Nature's History: Travellers and Travel Narratives in the Early Royal Society," *Annals of Science* 54 (1997): 269–292, esp. 281. Denis Leigh, "Medicine, the City and China," *Medical History* 18 (1974): 51–67.

26. Lisbet Koerner, *Linnaeus: Nature and Nation* (Cambridge, Mass.: Harvard University Press, 1999), 116–117, 136–139, 150–151.

27. J.-B. du Halde, *Description géographique, historique, chronologique, politique et physique . . . etc. etc.*, 4 vols. (Paris: P. G. Le Mercier, 1735).

28. Wilson and Caley, *Europe Studies China*, 13–14.

29. *Chinese Repository* 5 (1836–1837): 119. Postcolonial critics often emphasize the discursive liaisons between the objectifying vision of Europeans and a (gendered) Orient, between a desiring gaze and a (projected) seductively veiled body. There is ample evidence in the naturalists' writings to confirm this view. And few will deny that imperial imagination and aggressive cognition were inscribed in the discourse of nineteenth-century natural history. Our job here is not to dwell on these well-explored points, but to discover the strategies the naturalists developed to overcome the "obstacles" to their natural historical research.

30. *Chinese Repository* 5 (1836–37): 119.

31. Yen P'ing Hao, *The Comprador in Nineteenth-Century China: Bridge between East and West* (Cambridge, Mass.: Harvard University Press, 1970).

32. Coates, *The China Consul* (see Chapter 3, note 3), 81–86.

33. Ibid.; see also Stanley Wright, *Robert Hart and the Chinese Customs* (cited in Chapter 3, note 7), 277.

34. One can easily come up with a list of British missionary- and diplomat-sinologists: Robert Morrison, James Legge, Thomas Wade, Joseph Edkins, Herbert Giles, to name only the best known.

35. Linda Cooke Johnson, *Shanghai from Market Town to Treaty Ports, 1074–1858* (Stanford: Stanford University Press, 1995), chs. 7–12.

36. For an account of the early days of the Society, see *JNCB* 35 (1903–1904): i–xx.

37. E. C. Bridgman, "Inaugural Address," *Journal of the Shanghai Literary and Scientific Society* 1 (June 1858): 1–16. On p. 6. Bridgman was an American missionary.

38. There are as yet no general accounts of these journals. But see Frank King, *The China Coast Newspaper Project of the Center for Research Libraries and the Center for East Asian Studies* (Lawrence: University of Kansas Press, n.d.); it focuses on newspapers, but gives a general idea of the busy publishing scene in Western communities in China. Both the *Chinese Repository* and the *Chinese Recorder* were founded by American missionaries. On the *Chinese Repository*, see Murray A.

Rubinstein, "The Wars They Wanted: American Missionaries' Use of *The Chinese Repository* before the Opium War," *The American Neptune* 48, no. 1 (1988): 271–282.

39. British academic sinology never caught up with its continental counterpart. But the large British population in China and their facilities made possible the growth of a wide range of areas of interest, including natural history.

40. *Journal of the Shanghai Literary and Scientific Society* 1 (June 1858), "Preface." Emil Bretschneider made the same point in the preface to his *On the Study and Value of Chinese Botanical Works, with Notes on the History of Plants and Geographical Botany from Chinese Sources* (Foochow: Rozario, Marcal & Co., 1871). The work first appeared in several parts in *Chinese Recorder* 3 (1870).

41. W. F. Mayers, "On the Introduction of Maize into China," *The Pharmaceutical Journal and Transactions*, 3rd sr., 1 (1870–1871): 522–525.

42. For the general importance of nomenclature in nineteenth-century natural history, see Harriet Ritvo, *The Platypus and the Mermaid, and Other Figments of the Classifying Imagination* (Cambridge, Mass.: Harvard University Press, 1997), chs. 1 and 2.

43. Robert Swinhoe, "The Small Chinese Lark," *JNCB*, no. 3 (1859): 288.

44. Robert Swinhoe, "On a New Rat from Formosa," *Proc. Zool. Soc.* (1864): 185–187.

45. Swinhoe to Newton, 23 November 1869, in CUL. MSS. Alfred Newton Papers. Swinhoe writes, "Soo Tungpo is a good classical Chinese name, and Science might well admit such worthy names. However, as it offends you."

46. E. Bretschneider, "Botanicon Sinicum: Notes on Chinese Botany from Native and Western Sources," *JNCB*, n.s., 16 (1881): 18–230. On p. 110.

47. Heude's taxonomic work would, however, prove to be controversial because of his anti-Darwinian point of view. He was also an important scientific traveler and field naturalist in China. See *Mémoires concernant l'histoire naturelle de l'empire chinois*, tome v., second cahier (Shanghai: Impr. de la Mission catholique, 1906), 1–29; P. Fournier, *Voyages et découvertes scientifiques des missionnaires naturalistes français* (Paris: Paul Lechevalier, 1932), 36–42.

48. Fournier, *Voyages et découvertes scientifiques*, 67–91; Emmanuel Boutan, *Le nuage et la vitrine: Une vie de Monsieur David* (Bayonne: Raymond Chabaud, 1993) is a biography based on David's correspondence.

49. Peter Galison, *Image and Logic: A Material Culture of Microphysics* (Chicago: University of Chicago Press, 1997), ch. 1.

50. Similar issues have been discussed in a colonial context. See, for example, Satpal Sangwan, "From Gentlemen Amateurs to Professionals: Reassessing the Natural Science Tradition in Colonial India, 1780–1840," in Richard Grove, Vinita Damodaran, and Satpal Sangwan, eds., *Nature and the Orient* (see Chapter 1, note 8), 210–229; Reingold and Rothenberg, eds., *Scientific Colonialism: A Cross-Cultural Comparison* (see "Introduction," note 6).

51. There are as yet no adequate studies of the Chinese literature about their natural environment or the Chinese attitudes toward the living world. Joseph Needham, *Science and Civilization in China* (Cambridge, England: Cambridge University Press, 1954–), vol. 6, *Biology*, is the closest thing we have. But as its title indicates, the work is limited by Needham's tendency to impose modern scientific categories on Chinese knowledge traditions.

52. Theophilus Sampson, *Botanical and Other Writings on China, 1867–1870*, ed. H. Walravens (Hamburg: C. Bell Verlag, 1984), 19.

53. T. Watters, "Chinese Notions about Pigeons and Doves," *JNCB*, n.s., 4 (1867): 225–241; idem, "Chinese Fox-Myths," *JNCB*, n.s., 8 (1873): 47–49.

54. The British were very interested in Chinese medicine and medicinals. See Denis Leigh, "Medicine, the City and China"; Roberta E. Bivins, *Acupuncture, Expertise and Cross-Cultural Medicine* (Houndmills, Britain: Palgrave, 2000), which may be read together with Fa-ti Fan, "Science and Medicine, Asia and Europe," *Metascience* 11 (July 2002): 177–184.

55. Daniel Hanbury, "Notes on Chinese Materia Medica," collected in his *Science Papers* (London, 1876). He corresponded with Henry Hance, William Lockhart, and F. Porter Smith. The latter two were missionary doctors. See RPS: Hanbury Papers, P273 [8], [62]; Hanbury Miscellaneous Letters, P300 [39]; P301 [34], P313 [1], and so on. Most of the letters were replies to Hanbury's questions about certain Chinese drugs.

56. F. Porter Smith, *Contributions towards the Materia Medica and Natural History of China, for the Use of Medical Missionaries and Native Medical Students* (Shanghai: American Presbyterian Mission Press, 1871). There was also a potential audience in Europe, and Hanbury helped to find a London publisher for the work. RPS: Hanbury Miscellaneous Letters 1871, P302 [72], [73].

57. NBG: Augustine Henry Papers, 581.634, "Pharmac. Notes." Augustine Henry, "Vegetable Productions, Central China," *Bulletin of Miscellaneous Information Kew Gardens*, no. 33 (1889): 225–227; idem, *Notes on Economic Botany of China* (Kilkenny, Ireland: Boethius Press, 1986 [1893]) and "Chinese Drugs and Medicinal Plants," *Pharmaceutical Journal* 68 (1902): 316–319, 322–324.

58. Charles Ford, Ho Kai, and William Edward Crow, "Notes on Chinese Materia Medica," *China Review* 15 (1886–87): 214–220, 274–276, 345–347; 16 (1887–88): 1–19, 65–73, 137–161. Ford evidently took pride in this enterprise. See his letters to Thiselton-Dyer, Kew: Chinese and Japanese Letters, 150 (310), (322), (323), (356).

59. Janet Browne, *The Secular Ark: Studies in the History of Biogeography* (New Haven: Yale University Press, 1983); idem, "Biogeography and Empire," in Jardine, Secord, and Spray, eds., *Cultures of Natural History*, 305–321; Michael Dettelbach, "Humboldtian Science," in *Cultures of Natural History*, 287–304; Philip F. Rehbock, *The Philosophical Naturalists: Themes in Early Nineteenth-Century British Biology* (Madison: University of Wisconsin Press, 1983), part 2; Malcom Nicolson, "Alexander von Humboldt and the Geography of Vegetation," in Andrew Cunningham and Nicolas Jardine, eds., *Romanticism and the Sciences* (Cambridge, England: Cambridge University Press, 1990), 169–185.

60. For example, T. W. Kingsmill's presidential address, "Border Lands of Geology and History," *JNCB*, n.s., 11 (1877): 1–31.

61. Alphonse de Candolle, *Géographie botanique raisonnée, ou exposition des faits principaux et des lois concernant la distribution géographique des plantes de l'époque actuelle*, vol. 2 (Paris: Masson, 1855), ch. 9; idem, *Origin of Cultivated Plants*, ch. 2 (New York: Hafner Publishing Company, 1967 [1886]).

62. Alphonse de Candolle, *Géographie botanique raisonnée*, vol. 2, 979–980.

63. Bretschneider, *On the Study and Value of Chinese Botanical Works*, 6–7; idem, "Botanicon Sinicum," 20–21.

64. References are numerous and can be found in Berthold Laufer, *Sino-Iranica: Chinese Contributions to the History of Civilization in Ancient Iran, with Special Reference to the History of Cultivated Plants and Products* (Chicago: Field Museum, 1919). Most of the disagreements were about whether a particular plant was introduced by Zhang Qian. Few, if any, questioned the story that Zhang Qian brought back some exotic plants from central Asia.

65. Bretschneider, *On the Study and Value of Chinese Botanical Works*, 7.

66. Bretschneider, "Fu-sang, or Who Discovered America?" *Chinese Recorder* 3 (1870–71): 114–120; Sampson, "Buddhist Priests in America," in his *Botanical and Other Writings*, 30–31.

67. Anthony Pagden, *European Encounters with the New World* (New Haven: Yale University Press, 1993), ch. 1.

68. Georges Métailié, "La création lexicale dans le premier traité de botanique occidentale publié en chinois (1858)," *Documents pour l'histoire du vocabulaire scientifique* 2 (1981): 65–73; Pan Jixing, "Tan 'Zhiwuxue' yi ci zai Zhongguo he Riben de youlai," *Daziran tansuo* 3 (1984): 167–172; Zhongguo zhiwuxue hui, ed., *Zhongguo zhiwuxue shi* (Beijing: Kexue chubanshe, 1994), 122–123. The term *zhiwu* means the plant, and *xue* means an organized body of learning. *Xue* was frequently adopted by the translators to denote Western or Western-styled disciplines of learning. *Bowu* is "a wide range of things." The Chinese did have a genre called *bowu zhi*, "records of a wide range of things," whose catholic inclusion of natural things must have reminded the Western translators of Pliny's *Natural History* and other similar works, hence the rendition of natural history into *bowu xue*. However, *bowu* did not have the connotation that it referred only to natural objects until it was associated with *xue* in the neologism. Similarly, *zhiwu* was a traditional term. It had been used to denote the plant, for example, in Wu Qijun's *Zhiwu mingshi tukao* (1848), which might be translated as "the pictorial study of the names and natures of plants."

69. See Chapter 1. See also Jack Goody, *The Cultures of Flowers* (Cambridge, England: Cambridge University Press, 1993); Craig Clunas, *Fruitful Sites: Garden Culture in Ming Dynasty China* (London: Reaktion Books, 1996); Wang Yi, *Yuanlin yu Zhongguo wenhua* (Shanghai: Renmin chubanshe, 1990).

70. O. F. von Möllendorff, "The Vertebrata of the Province of Chili with Notes on Chinese Zoological Nomenclature," *JNCB*, n.s., 11 (1877): 41–111. On p. 44. For a concise introduction to *Bencao gangmu*, see Paul Unschuld, *Medicine in China: A History of Pharmaceutics* (Berkeley: University of California Press, 1986), 145–164; Joseph Needham, *Botany*, vol. 6, part 1, *Science and Civilization in China* (Cambridge, England: Cambridge University Press, 1986), 308–321. For Li Shizhen, see Nathan Sivin's essay in *Dictionary of Scientific Biography*, vol. 8, 390–398. See also Zhongguo zhiwu xuehui, ed., *Zhongguo zhiwuxue shi*, 69–81; Zhongguo yaoxuehui, *Li Shizhen yanjiu lunwen ji* (Wuhan: Hubei kexue chubanshe, 1985); Georges Métailié, "Des plantes et des mots dans le *Bencao gangmu* de Li Shizhen," *Extrême-Orient, Extrême-Occident* 10 (1988): 27–43. Pan Jixing discusses the transmission of the *Bencao* to Europe in his *Zhongwai kexue zhi jiaoliu*, 206–214.

71. S. Wells Williams, *The Middle Kingdom*, vol. 1 (New York: Charles Scribner's Sons, 1913 [1882]), 372.

72. Ibid., 370.

73. Möllendorff, "The Vertebrata," 44–45. For an interesting comparative study of *Bencao* and a Renaissance herbal, see Georges Métailié, "Histoire naturelle

et humanisme en Chine et en Europe au XVIᵉ siècle," *Revue d'histoire des sciences* XLII/4 (1989): 353-374. He argues that there are significant similarities between the botanical knowledge in the two works, though one can probably question his grounds for comparison.

74. Möllendorff, "The Vertebrata," 42.
75. Needham, *Botany*, 22-23.
76. Bretschneider, "Botanicon Sinicum," 66.
77. Ibid., 66-67.
78. Ibid., 65.
79. Ibid., 65-66.
80. Richard Rudolph, "Illustrated Botanical Works in China and Japan," in Thomas Buckman, ed., *Bibliography and Natural History* (Lawrence: University of Kansas Press, 1966); André Georges Haudricourt and Georges Métailié, "De l'illustration botanique en Chine," *Études chinoises* 13 (1994): 381-416.
81. *JNCB*, n.s., 25 (1890-1891): 403.
82. Bretschneider, "Botanicon Sinicum," 50. Haudricourt and Métailié, "De l'illustration botanique en Chine" compares Chinese and Renaissance European herbals and argues that the representations of plants in Chinese herbals, such as *Bencao*, remained fundamentally verbal. It is a point well taken, though one still wonders why the works included plates at all.
83. Bretschneider, "Botanicon Sinicum," 55.
84. Sampson, *Botanical and Other Writings*, 41.
85. A. A. Fauvel, "Alligators in China," *JNCB*, n.s., 13 (1878): 1-36.
86. Robert Swinhoe to Richard Owen, 18 February 1870, NHML: Owen Correspondence, vol. 25, ff. 69-70. Möllendorff, "The Vertebrata," 44.
87. Bretschneider, "Botanicon Sinicum," 35.
88. Both Georges Métailié and modern Chinese scholars convincingly argue that Wu Qijun had been influenced by the *kaozheng* philological methodology popular among Chinese scholars since the late eighteenth century. The *kaozheng* school emphasized dogged evidential scholarship. Generically speaking, Wu's *Zhiwu* was more a pictorial dictionary of plants than an herbal. Henan sheng kexue jishu xiehui, ed., *Wu Qijun yanjiu* (Zhenzhou: Zhongzhou guji chubanshe, 1991), 55-57; André Georges Haudricourt and Georges Métailié, "De l'illustration botanique en Chine." For the *kaozheng* school, see Benjamin Elman, *From Philosophy to Philology: Intellectual and Social Aspects of Change in Late Imperial China* (Cambridge, Mass.: Harvard University Press, 1984).
89. Bretschneider's *On the Study and Value of Chinese Botanical Works* includes eight plates from *Zhiwu mingshi tukao*.
90. Sampson, *Botanical and Other Writings*, 32.
91. Henry to William Thiselton-Dyer, 13 May 1887. Kew: Chinese and Japanese Letters 151 (604).
92. For example, E. C. Baber, *Travels and Researches in Western China*, Supplementary Papers of the Royal Geographical Society 1 (1882); idem, *Report by Mr. Baber on the Route Followed by Mr. Governor's Mission between Talifu and Momein*, Parliamentary Papers, China No. 3 (1878); *Report by Mr F. S. A. Bourne of a Journey in South-Western China*, Parliamentary Papers, China No. 1 (1888); *Report by Mr. Hosie of a Journey through the Provinces of Suu-ch'uan, Yunnan, and Kuei Chou: February 11 to June 14, 1883*, Parliamentary Papers, China no. 2 (1884).
93. See, for example, Robert Swinhoe, "On the Mammals of the Island of

Formosa (China)," *Proc. Zool. Soc.* (1862): 347–365; "The Ornithology of Formosa, or Taiwan," *Ibis* 5 (1863): 198–219, 250–311, 377–435; "On the Mammals of Hainan," *Proc. Zool. Soc.* (1870): 224–236.

94. Robert Swinhoe, "Neau-Show," *JNCB*, n.s., 2(1865): 39–52.

95. Möllendorff, "The Vertebrata," 46. See also his "Trouts in China," *China Review* 7 (July 1878–June 1879), 276–278.

96. Bretschneider, "Botanicon Sinicum," 87.

97. Sampson, *Botanical and Other Writings*, 31–36; Bretschneider, "Botanicon Sinicum," 92–95.

98. Alfred Newton, "Abstract of Mr. J. Wolley's Researches in Iceland respecting the Gare-fowl or the Great Auk (*Alca impennis*, Linn.)," *Ibis* 3 (1861): 375–399; Symington Grieve, *The Great Auk, or Garefowl (Alca impennis, Linn.): Its History, Archaeology, and Remains* (London: T. C. Jack [etc.], 1885).

99. Paul L. Farber, "The Type Concept in Zoology during the First Half of the Nineteenth Century," *Journal of the History of Biology* 11 (1976): 93–119.

100. Hance to Henry, 1 April 1885, 7 June 1885, in NBG: Letters to Henry. (The letters are not numbered.)

101. SOAS: Bowra Papers, MS. English. 201813, Box 2, No. 7, Bowra's 1863 diary, July 18.

102. Bretschneider, "Botanicon Sinicum," 19.

103. Ibid., 67.

104. Michael Adas, *Machines as the Measure of Men* (Ithaca: Cornell University Press, 1989); J. W. Burrow, *Evolution and Society: A Study in Victorian Social Theory* (Cambridge, England: Cambridge University Press, 1966), 11–14; George Stocking, Jr., *Victorian Anthropology* (New York: Free Press, 1987), 174–175; Robert Nisbet, *Social Change and History: Aspects of the Western Theory of Development* (Oxford: Oxford University Press, 1969), 189–208.

105. W. H. Medhurst, *A Dissertation on the Theology of the Chinese* (Shanghai, 1847); Arthur Wright, "The Chinese Language and Foreign Ideas," in *Studies in Chinese Thought*, ed. Arthur Wright (Chicago: University of Chicago Press, 1953), 286–303; Gernet, *China and the Christian Impact*, 238–247.

106. "The Advisability, or the Reverse, of Endeavouring to Convey Western Knowledge to the Chinese through the Medium of Their Language," *JNCB*, n.s., 21 (1886): 1–21. Some of the opinions resembled to some extent the controversy about education in India earlier in the century. See, for example, Adas, *Machines as the Measure of Men*, 271–292. The uniqueness of the Chinese language, especially its many ideograms, presented new challenges to Western educators in China.

107. Tejaswini Niranjana, *Siting Translation: History, Post-Colonialism, and the Colonial Context* (Berkeley: University of California Press, 1992) discusses some theoretical issues about power and translation.

108. Bretschneider, "Botanicon Sinicum," 21.

109. On "trust" in science, see, for example, Steven Shapin, *A Social History of Truth*, esp. 243–266.

110. Robert Swinhoe to Richard Owen, 18 February 1870, NHML: Owen Correspondence, vol. 25, ff. 69–70.

111. Richard Owen, "On Fossil Remains of Mammals Found in China," *Quarterly Journal of the Geological Society of London* 26 (1870): 417–434.

112. Hance's introductory remarks to W. F. Mayers, "On the Introduction of

Maize into China," *The Pharmaceutical Journal and Transactions*, 3rd sr., 1 (1870–1871): 522–525.

113. Sampson, *Botanical and Other Writings*, 17.

114. Harriet Ritvo, "Zoological Nomenclature and the Empire of Victorian Science," in Lightman, ed., *Victorian Science in Context*, 334–353.

115. A. A. Fauvel, "Alligators in China," *JNCB*, n.s., 13 (1878): 1–36.

116. On the introduction of exotic animals and plants into China, see Berthold Laufer, *Sino-Iranica*; S. A. M. Adshead, *China in World History*, 2nd ed. (London: Macmillan, 1995); Edward Schafer, *The Golden Peaches of Samarkand: A Study of T'ang Exotics* (Berkeley: University of California Press, 1963); Shiu Ying Hu, "History of the Introduction of Exotic Elements into Traditional Chinese Medicine," *Journal of the Arnold Arboretum* 71 (1990): 487–526.

117. Sampson, *Botanical and Other Writings*, 23.

118. Bretschneider, "Botanicon Sinicum," 66; Ernst Faber, "Contribution to the Nomenclature of Chinese Plants," *JNCB* 38 (1907): 97–164.

119. Augustine Henry, "Chinese Names of Plants," *JNCB*, n.s., 22 (1887): 233–283.

120. Henry to H. B. Morse, 17 June 1893, Kew: A. Henry Letters to H. B. Morse, 3–5.

121. Fauvel, "Alligators in China," 4.

122. Ibid., 4–5.

123. W. F. Mayers, "The Mammoth in Chinese Records," *China Review* 6 (July 1877–June 1878): 273–276.

124. Candolle, *Origin of Cultivated Plants*, passim.

125. Frederick Burkhardt et al., eds., *The Correspondence of Charles Darwin* (Cambridge, England: Cambridge University Press, 1989), vol. 15, 448.

126. Ibid., 510–511. A third of the people he asked for assistance were located in China: Fortune, who was then collecting tea plants in China, J. C. Bowring in Hong Kong, Swinhoe and Harry Smith Parkes, both of the Consular Service, the American missionary doctor Daniel Jerome Macgowan of Ningbo, William Aurelius Harland, a surgeon in Hong Kong, and the missionary doctor William Lockhart of Shanghai.

127. Charles Darwin, *The Descent of Man, and Selection in Relation to Sex*, vol. 2 (New York: D. Appleton & Co., 1871), 16. W. F. Mayers, "Gold Fish Cultivation," *Notes and Queries on China and Japan* 2 (1868): 123–124. Pan Jixing has painstakingly identified the original Chinese texts of the almost one hundred references to Chinese sources in Darwin's major books. See his *Zhongwai kexue zhi jiaoliu*, ch. 1

5. Travel and Fieldwork in the Interior

1. Scholars have used "experience" as an interpretive concept in their studies of social drama and of human relationships with their environment, and I have found their work inspiring in some respects. For example, Victor W. Turner and Edward M. Bruner, eds., *The Anthropology of Experience* (Urbana: University of Illinois Press, 1986); Yi-Fu Tuan, *Space and Place: The Perspective of Experience* (Minneapolis: University of Minnesota Press, 1977); idem, *Passing Strange and Wonderful: Aesthetics, Nature, and Culture* (Washington, D.C.: Island Press, 1993). On the body and senses in science, see, for example, Gillian Beer, "Four Bodies on the 'Beagle':

Touch, Sight and Writing in a Darwin Letter," in her *Open Fields: Science in Cultural Encounters* (Oxford: Clarendon Press, 1996), 13–30; Christopher Lawrence and Steven Shapin, eds., *Science Incarnate: Historical Embodiments of Natural Knowledge* (Chicago: University of Chicago Press, 1998). On aesthetics and natural history, see, for example, Michael Shortland, "Darkness Visible: Underground Culture in the Golden Age of Geology," *History of Science* 32 (1994): 1–61; Lynn Merrill, *The Romance of Victorian Natural History* (see "Introduction," note 4); Barbara T. Gates, *Kindred Nature* (see Chapter 3, note 50).

2. There is a growing interest among historians of science in field science and fieldwork. See, for example, Jane Camerini, "Remains of the Day: Early Victorians in the Field," in Lightman, ed., *Victorian Science in Context* (see Chapter 2, note 4), 354–377; Henrika Kuklick and Robert E. Kholer, eds., *Science in the Field*, Osiris (1996), vol. 11; Alex Soojung-Kim Pang, "The Social Event of the Season: Solar Eclipse Expeditions and Victorian Culture," *Isis* 84 (1993): 252–277; Robert E. Kholer, "Place and Practice in Field Biology," *History of Science* 40 (2002): 189–210; idem, *Landscapes and Labscapes: Exploring the Lab-Field Frontier in Biology* (Chicago: University of Chicago Press, 2002); Bruno Latour, "Circulating Reference: Sampling the Soil in the Amazon Forest," in his *Pandora's Hope: Essays on the Reality of Science Studies* (Cambridge, Mass.: Harvard University Press, 1999), 24–79; David Oldroyd, *The Highlands Controversy: Constructing Geological Knowledge through Fieldwork in Nineteenth-Century Britain* (Chicago: University of Chicago Press, 1990); Anne Larsen, "Equipment for the Field," in Jardine, Secord, and Spray, eds., *Cultures of Natural History* (see "Introduction," note 4), 358–377; idem, "Not Since Noah: English Scientific Zoologists and the Craft of Collecting, 1800–1840" (Ph.D. diss., Princeton University, 1993); Barbara and Richard Mearns, *The Bird Collectors* (San Diego: Academic Press, 1997), ch. 3. George W. Stocking, Jr., ed., *Observers Observed: Essays on Ethnographic Fieldwork* (Madison: University of Wisconsin Press, 1983) focuses on ethnography.

3. The literature on scientific travel and exploration is considerable. For example, Mary Louise Pratt, *The Imperial Eye: Travel Writing and Transculturation* (London: Routledge, 1992); Clare Lloyd, *The Traveling Naturalists* (Seattle: University of Washington Press, 1985); Peter Raby, *Bright Paradise: Victorian Scientific Travelers* (Princeton: Princeton University Press, 1996); Anthony Pagden, *European Encounters with the New World: From Renaissance to Romanticism* (New Haven: Yale University Press, 1993); Gillian Beer, "Traveling the Other Way," in Jardine, Secord, and Spray, eds., *Cultures of Natural History*, 322–337; Paul Carter, *The Road to Botany Bay: An Exploration of Landscape and History* (Chicago: University of Chicago Press, 1989); Bernard Smith, *European Vision and the South Pacific*, 2nd ed. (New Haven: Yale University Press, 1988). Beau Riffenburgh, *The Myth of the Explorer: The Press, Sensationalism, and Geographic Discovery* (Oxford: Oxford University Press, 1994); Barbara M. Stafford, *Voyage into Substance: Art, Science, Nature, and the Illustrated Travel Account, 1760–1840* (Cambridge, Mass.: MIT Press, 1984); George Stocking, *Victorian Anthropology* (New York: Free Press, 1987), ch. 3.

4. Accounts of these events can be found in Bretschneider, *History*; Numa Broc, "Les explorateurs français du XIXe siècle reconsidérés," *Rev. Franç. d'Hist. d'Outre-Mer*, LXIX (1982): 237–273; idem, "Les voyageurs français et la connaissance de la Chine (1860–1914)," *Revue historique* 276, no. 1 (1986): 85–131; P. Fournier, *Voyages et découvertes scientifiques des missionaires naturalistes français*

(Paris: Paul Lechevalier, 1932), part II; Barbara and Richard Mearns, *The Bird Collectors*, 266–283.

5. Evelyn Rawski presents an overview of current literature and issues along these lines in her "Reenvisioning the Qing: The Significance of the Qing Period in Chinese History," *Journal of Asian Studies* 55 (1996): 829–850.

6. There has been an explosion of interest in Chinese travel in the past few years, but much of the work is still in progress. See, for example, Richard E. Strassberg, ed., trans., *Inscribed Landscapes: Travel Writing from Imperial China* (Berkeley: University of California Press, 1994); Timothy Brook, "Communications and Commerce," in Denis Twitchett and Frederick W. Mote, eds., *The Cambridge History of China*, vol. 8, part II (Cambridge, England: Cambridge University Press, 1998), 579–707; Hu Ying, "Reconfiguring Nei/Wai: Writing the Woman Traveler in the Late Qing," *Late Imperial China* 18 (1997): 72–99; Wang Liping, "Paradise for Sale: Urban Space and Tourism in the Social Transformation of Hangzhou, 1589–1937" (Ph.D. diss., University of California, San Diego, 1997); Ho-chin Yang, "China's Routes to Tibet during the Early Qing Dynasty: A Study of Travel Accounts" (Ph.D. diss., University of Washington, 1994); Emma Teng, "Travel Writing and Colonial Collecting: Chinese Travel Accounts of Taiwan from the Seventeenth through Nineteenth Centuries" (Ph.D. diss., Harvard University, 1997); Susan Naquin and Chün-fang Yü, eds., *Pilgrims and Sacred Sites in China* (Berkeley: University of California Press, 1992). Arthur Waley, *Yuan Mei: Eighteenth Century Chinese Poet* (Stanford: Stanford University Press, 1956) vividly describes the famous writer Yuan Mei's interesting travels.

7. On the exploration of Africa, see Frank McLynn, *Hearts of Darkness: The European Exploration of Africa* (New York: Carrol & Graft, 1992); Johannes Fabian, *Out of Our Minds: Reason and Madness in the Exploration of Central Africa* (Berkeley: University of California Press, 2000). Peter Raby, *Bright Paradise*, chs. 3 and 4, deals with the exploration of South America. Henry Walter Bates, *The Naturalist on the River Amazons: A Record of Adventures, Habits of Animals, Sketches of Brazilian and Indian Life, and Aspects of Nature under the Equator, during Eleven Years of Travel* (New York: Penguin Books, 1989 [1863]).

8. The British in China made a survey of inland communications in China in 1890, and the results may be taken as a general profile of their knowledge and opinions about inland travel in China. "Inland Communications in China," *JNCB*, n.s., 28 (1893–1894): 1–213.

9. For example, *Report by Mr. Hosie of a Journey through the Provinces of Ssuch'uan, Yunnan, and Kuei Chou, February 11 to June 14, 1883*. Parliamentary Papers, China. No. 2 (1884), 52.

10. It may be worth noting that the traveling naturalists typically compared Chinese roads and inns (unfavorably) with those in their homelands, not with struggling through and sleeping in a rainforest or desert.

11. Sybille Fritzsche, "Narrating China: Western Travelers in the Middle Kingdom after the Opium War" (Ph.D. diss., University of Chicago, 1995) analyzes the writings of Western travelers in China. Joshua Vogel, *The Literature of Travel in the Japanese Rediscovery of China, 1862–1945* (Stanford: Stanford University Press, 1996) focuses on Japanese travelers in twentieth-century China.

12. Gerald Graham, *The China Station: War and Diplomacy, 1830–1860* (New York: Oxford University Press, 1978), esp. 254–275, 407–421.

13. Adrian Desmond, *Huxley: The Devil's Disciple* (London: Michael Joseph,

1994), 53–146; Helen Rozwadowski, "Fathoming the Ocean: Discovery and Exploration of the Deep Sea, 1840–1880" (Ph.D. diss., University of Pennsylvania, 1996); idem, "Small World: Forging a Scientific Maritime Culture for Oceanography," *Isis* 87 (1996): 409–429. See also Matthew Robert Goodrum, "The British Sea-Side Studies, 1820–1860: Marine Invertebrates, the Practice of Natural History, and the Depiction of Life in the Sea" (Ph.D. diss., Indiana University, Bloomington, 1997).

14. Arthur Adams, ed., *The Zoology of the Voyage of H. M. S. Samarang; under the Command of Captain Sir Edward Belcher* . . . (London: Reeve and Benham, 1850). See also his *Notes from a Journal of Research into the Natural History of the Countries Visited during the H. M. S. Samarang, under the Command of Captain Sir Edward Belcher* (London: Reeve, Benham, and Reeve, 1848). In addition, Adams published frequently on marine animals in the China and Japan seas in *The Annals and Magazine of Natural History* in the 1840s through the 1860s.

15. Cuthbert Collingwood, *Rambles of a Naturalist on the Shores and Waters of the China Sea* (London: John Murray, 1868). He also published many articles on his discoveries during the voyage in *The Annals and Magazine of Natural History* in the late 1860s. On Collingwood, see Nora McMillan's biographic sketch of him in *The Linnean* 18 (April 2001). P. W. Bassett-Smith, *China Sea. Report on the Results of Dredgings Obtained on the Macclesfield Bank in H. M. S. "Rambler"* . . . (London: Her Majesty's Stationery Office, 1894). See also NHML: Z. Keeper's Archives. 1.41. Letters. 1892. Jan.–June. No. 18–19 and 1.42. 1892. July–Dec. No. 20–21.

16. On plant hunting in China, see Tyler Whittle, *The Plant Hunters* (New York: PAJ Publications, 1988); Kenneth Lemmon, *The Golden Age of Plant Hunters* (New York: Barnes, 1969), ch. 5 (which focuses on William Kerr); E. H. M. Cox, *Plant Hunting in China* (London: Collins, 1945); Alice Coats, *The Plant Hunters* (New York: McGraw-Hill, 1969), 86–140; Stephen Spongberg, *A Reunion of Trees: The Discoveries of Exotic Plants and Their Introduction into North American and European Landscapes* (Cambridge, Mass.: Harvard University Press, 1990), chs. 3–6.

17. On hunting and empire, see John MacKenzie, *The Empire of Nature: Hunting, Conservation and British Imperialism* (Manchester: Manchester University Press, 1988); Harriet Ritvo, *The Animal Estate: The English and Other Victorian Creatures in the Victorian Age* (Cambridge, Mass.: Harvard University Press, 1987), ch. 6; William Beinart, "Empire, Hunting and Ecological Change in Southern and Central Africa," *Past and Present* 42 (1990): 402–436; M. S. S. Pandian, "Hunting and Colonialism in the Nineteenth-Century Nilgiri Hills of South India," in Grove, Damodaran, and Sangwan, eds., *Nature and the Orient* (see Chapter 1, note 8), 273–298. Hunting, collecting, imperialism, and the modern zoo were closely connected during this period, and they were not limited to British imperialism. See Nigel Rothfels, *Savage and Beasts: The Birth of the Modern Zoo* (Baltimore: Johns Hopkins University Press, 2002); Elizabeth Hanson, *Animal Attractions: Nature on Display in American Zoos* (Princeton: Princeton University Press, 2002).

18. Robert Stafford, *Scientist of Empire: Sir Roderick Murchison, Scientific Exploration and Victorian Imperialism* (Cambridge, England: Cambridge University Press, 1989), 132–143, discusses Murchison's connections with the British in China. For the importance of fieldwork to geology, see Oldroyd, *The Highlands Controversy*; Martin Rudwick, *The Great Devonian Controversy: The Shaping of Scientific Knowledge among Gentlemanly Specialists* (Chicago: University of Chicago Press, 1985), 37–41.

19. T. W. Kingsmill, "Freiherr Ferdinand von Richthofen," *JNCB*, n.s., 37 (1906): 218–220; Wang Gen Yuan and Michel Gert, "The German Scholar Ferdinand von Richthofen and Geology in China," in *Interchange of Geoscience Ideas between the East and the West*," ed. Wang et al. (see Chapter 3, note 109), 47–54; David Oldroyd and Yang Jing-Yi, "On Being the First Western Geologist in China: The Work of Raphael Pumpelly (1837–1923)," *Annals of Science* 53 (1996): 107–136.

20. A. A. Fauvel, "Shanghai Museum, Report of the Curator for the Year 1878," *JNCB*, n.s., 13 (1878): xvi. In his presidential address, T. W. Kingsmill, a geologist, urged the members of the NCBRAS to study geology, but apparently few rose to his call. See his "Borderland of Geology and History," *JNCB*, n.s., 11 (1877): 6–31.

21. Fletcher, *The Story of the Royal Horticultural Society*, 148–152.

22. William Gardener, "Robert Fortune and the Cultivation of Tea in the United States," *Arnoldia* 31 (1971): 1–18; idem, "Robert Fortune: His Search for Tea Plants," *Gardeners Chronicle Gardening Illustrated*, 21 December 1963; idem, "Robert Fortune: A Plant Hunter of Resource," *Gardeners Chronicle Gardening Illustrated*, 14 September 1963, 185–186. Fortune's travel books were based on his correspondence to popular magazines in England, such as the *Gardeners Chronicle*, and reviews of his books were generally enthusiastic. For example, *Edinburgh Review* 88 (1848): 403–429; *The Athenaeum*, no. 1014 (1847): 358–361; no. 1015 (1847): 386–388; no. 1279 (1852): 481–483; no. 1545 (1857): 717–718; *Quarterly Review* 102 (1857): 126–165.

23. For Kew collectors during this period, see Desmond, *Kew: The History of the Royal Botanical Gardens* (see Chapter 3, note 28), 206–222. Bretschneider, *History*, 539–544, 682–688; Coats, *The Plant Hunters*, 76–78.

24. Robert Fortune, *Yedo and Peking: A Narrative of a Journey to the Capitals of Japan and China* (London: John Murray, 1863); James H. Veitch, *Hortus Veitchii: A History of the Rise and Progress of the Nurseries of Messrs. James Veitch and Sons . . .* (London: James Veitch & Sons, 1906), 49–52, 79–83.

25. Bretschneider, *History*, 774–794; Cox, *Plant Hunting in China*, 106–110, 136–169; Coats, *The Plant Hunters*, 116–127.

26. Robert Fortune to John Lindley, 10 August 1844, RHS: Robert Fortune Correspondence. 69D16; Kew: Kew Collectors, vol. 8. Wilford, (50); Kew: Kew Collectors, vol. 9. Oldham, (85–88).

27. For example, Robert Fortune was told, "To all collections of living plants & seeds the Society lays exclusive claim. . . . But any other collections which you may form will be your private property." See "Instructions to Mr Robert Fortune proceeding to China in the service of the Horticultural Society of London," in RHS: 69D15. Kew had similar restrictions about their collectors' exertions. For example, in Richard Oldham's contract, it is stated that "on no account is Richard Oldham to give any part of his Collections to any private individual, or to assist such in making Collections." Kew: Kew Collectors, vol. 9. Oldham, (6).

28. John Lindley (?) to Fortune, 6 March 1845, RHS: 69D15. His emphasis.

29. Ray Desmond, *Kew: The History of the Royal Botanical Gardens*, 101–102.

30. Kew: Kew Collectors, vol. 4. Oldham, (28–29).

31. Kew: Kew Collectors, vol. 4. Oldham, (29–31), (36–37), (37–42).

32. Kew: Kew Collectors, Wilford, (76), (77–78), (118), (121), (125), (138), (141), (150).

33. Katherine Bruner, John Fairbank, and Richard Smith, eds., *Entering China's Service: Robert Hart's Journals, 1854–1863* (Cambridge, Mass.: Harvard University Press, 1986), 30.

34. John Ward to William Hooker, 26 April 1859, Kew: Kew Collectors. Wilford (80); Henry Hance to William Hooker, 9 September 1863, Kew: Chinese and Japanese Letters, 57 (77).

35. Kew: Kew Collectors, vol. 4. Oldham, (33–36); Kew: Chinese and Japanese Letters, 151 (949), (949a), (953). See also Kew: Chinese and Japanese Letters, 57 (79).

36. Henry Hance to William Hooker, 27 May 1863, Kew: Chinese and Japanese Letters, 57 (76), (77); Kew: Kew Collectors, vol. 4. Oldham, (19–31).

37. Kew: Kew Collectors, vol. 4. Oldham, (5–6), (7–9).

38. A sportsman remarked that Chinese New Year was "a regular carnival for sportsmen," particularly because "the fields are free from laborers" who were "spending the holidays with their families." *China Review* 5 (July 1876–June 1877): 286–287.

39. Sybille Fritzche, "Narrating China," ch. 7; Lyman P. van Slyke, *Yangtze: Nature, History, and the River* (Reading, Mass.: Addison-Wesley Publishing Co., 1988) provides useful background information (esp. chs. 3, 7, and 8) and has a chapter on Western travelers on the Yangzi.

40. "Che-Foo in China," *Supplement* to the *Illustrated London News*, 9 December 1876, 561. Chefoo, a treaty port, was of course a seaside resort rather than a *hill* station. On the European fascination for the seaside, see Alain Corbin, *The Lure of the Sea: The Discovery of the Seaside in the Western World, 1750–1840* (Berkeley: University of California Press, 1994). On hill stations in India, see Dane Kennedy, *The Magic Mountains: Hill Stations and the British Raj* (Berkeley: University of California Press, 1996).

41. See, for example, *Illustrated London News* (1873): 256, 287, 476, 603–604, 622; (1875): 250, 279, 280; (1891): 765, 760, 777, 796, and numerous other places. The entomological collector A. E. Pratt also published a piece about his expedition in western China in *Illustrated London News* (1891): 511–512, 536, 577.

42. Isabella Bird, *The Yangtze Valley and Beyond. An Account of Journeys in China, Chiefly in the Province of Sze Chuan and Among the Man-Tze of the Somo Territory* (London: John Murray, 1899).

43. H. E. M. James, *The Long White Mountain, or a Journey in Manchuria* (London: Longmans, Green, and Co., 1888). As a hunting trip, this semiprivate expedition was unsuccessful. They bagged no big game. But it was also a scientific exploration connected with the Great Game. Patrick French, *Younghusband: The Last Great Imperial Adventurer* (London: Flamingo, 1995), 34–44.

44. One British sportsman in China told his fellow nimrods in Britain: "When I say shooting [in China] I do not mean the kind of sport to which one is accustomed at home, where to trespass a few yards on the grounds of another man will probably result in legal proceedings. . . . No, I mean where one can look on the whole empire of China and say, 'Here is my ground, here I can take my gun and my dogs and go just wherever, and do whatever, I please, without let or hindrance; shoot what I will, stay as long as I like without asking anyone's leave, and where keepers and game licences are unknown.'" An exaggeration to be sure, as are so many travelers' tales. But like many travelers' tales, too, it contains elements of

truth and it tells more than was intended. A modern reader is more likely to be struck by its blatant combination of sport and imperialism than to share the sportsman's fancy for hunting unlimited. The quote is from Oliver G. Ready, *Life and Sport in China*, 2nd ed. (London: Chapman & Hall, 1904), 46.

45. Ready, *Life and Sport in China*, 50; A. R. Margary, *The Journey of Augustus Margary* (London: Macmillan and Co., 1876), 54. H. T. Wade, ed., *With Boat and Gun in the Yangtze Valley* (Shanghai: Shanghai Mercury Office, 1895), ch. 1, describes the major shooting resorts in the Yangzi region. Jonathan Spence, *God's Chinese Son: The Taiping Heavenly Kingdom of Hong Xiuquan* (London: HarperCollins, 1996) describes in vivid detail the devastating battles between the Taipings and their enemies.

46. T. R. Jernigan, *Shooting in China* (Shanghai: Methodist Publishing House, 1908), 65–66; Wade, ed., *With Boat and Gun in the Yangtze Valley*, 16–17.

47. Wade, ed., *With Boat and Gun in the Yangtze Valley*, 156–164.

48. For example, William Spencer Percival, *The Land of the Dragon: My Boating and Shooting Excursions to the Gorges of the Upper Yangtze* (London: Hurst and Blackett, 1889). See also the books by Jernigan (note 46) and Wade (note 45).

49. On Chinese hunters, see note 68. On the concerns of the diminishing of game in the lower Yangzi, see Henling Thomas Wade's "Preface" in the second edition of his *With Boat and Gun in the Yangtze Valley* (Shanghai, 1910), iii–v. On hunting and conservation in other parts of the world, see, for example, David Elliston Allen, *The Naturalist in Britain* (see "Introduction," note 4), ch. 10; Mark V. Barrow, *A Passion for Birds: American Ornithology after Audubon* (Princeton: Princeton University Press, 1998); John MacKenzie, *The Empire of Nature*, chs. 8–11.

50. For basic biographic information on the three ornithologists, see *Ibis*, sr. 13, 5 (1935): 210, 889–890; Barbara and Richard Mearns, *The Bird Collectors*, 140–141.

51. For example, the curators' reports in *JNCB*, n.s., 11 (1877): viii–ix; 13 (1878): xiii–xviii; 15 (1880): xxii–xxv; 16, part II (1881): x–xv.

52. For example, the list of contributions during the year 1881 in *JNCB*, n.s., 16, part II (1881): xii–xv. It was common for the British (and the French) to establish natural history museums in the colonies and other settlements abroad, especially during the late nineteenth century. Susan Sheets-Pyenson, *Cathedrals of Science: The Development of Colonial Natural History Museums during the Late Nineteenth Century* (Kingston, Canada: McGill-Queen's University, 1988). For a postcolonial study of museums in colonial India, see Gyan Prakash, "Science 'Gone Native' in Colonial India," *Representations*, no. 40 (Fall 1992): 153–178. The British tried to set up a museum in Macao in the days of Old Canton (see Chapter 1), had one in Hong Kong, and established another in Shanghai. On the Hong Kong Museum, see *China Review* 5 (July 1876–June 1877): 72.

53. F. W. Styan, "On a Collection of Birds from Foochow," *Ibis*, sr. 5, 5 (1887): 215. The birds were from La Touche. La Touche describes his collecting ground in his "On Birds Collected and Observed in the Vicinity of Foochow and Swatow in South-Eastern China," *Ibis*, sr. 6, 4 (1892): 400–402. C. B. Rickett describes his collecting ground in his "On some Birds Collected in the Vicinity of Foochow," *Ibis*, sr. 6, 6 (1894): 215–216.

54. Bretschneider, *History*, 802–805; Cox, *Plant-Hunting in China*, 110–123, 132–135. For French explorations of China in the second half of the nineteenth century, see Numa Broc, "Les voyageurs français et la connaissance de la Chine

(1860–1914)," *Revue historique* 276, no. 1 (1986): 85–131; P. Fournier, *Voyages et découvertes scientifiques des missionnaires naturalistes français* (Paris: Paul Lechevalier, 1932), part II; Emmanuel Boutan, *Le nuage et la vitrine: Une vie de Monsieur David* (Bayonne: Raymond Chabaud, 1993).

55. Fortune, *A Journey to the Tea Countries of China*, 33. It was not uncommon for Western travelers in China to disguise themselves this way, especially when they intended to sneak out of the areas open to Westerners specified in the treaty articles. Another related reason was that it would reduce the attention from the natives. W. H. Medhurst, Sr., T. T. Cooper, and A. E. Pratt were among those who traveled in Chinese dress. See W. H. Medhurst, *A Glance at the Interior of China, Obtained during a Journey through the Silk and Green Tea Districts* (Shanghae [Shanghai]: Printed at the Mission Press, 1849), 1–11, 35–36, which gives a detailed description of his makeover and Chinese outfit. T. T. Cooper, *Travels of a Pioneer of Commerce in Pigtail and Petticoats; or an Overland Journey from China towards India* (London: John Murray, 1871); A. E. Pratt, *To the Snows of Tibet through China* (London: Longmans, Green, & Co., 1892).

56. On this mode of travel, see, for example, Medhurst, *A Glance at the Interior of China*, 29–34, 36–38; Pratt, *To the Snows of Tibet through China*, 41; G. E. Morrison, *An Australian in China* (Sydney: Augus and Robertson, 1972 [1895]), 50, 58–60, 89–91, 115; E. H. Wilson, *A Naturalist in Western China with Vasculum, Camera, and Gun. Being Some Account of Eleven Years' Travel, Exploration, and Observation in the More Remote Parts of the Flowery Kingdom*, 2 vols. in one (London: Cadogan Books, 1986 [1913]), vol. 1, 22–27. See also "Inland Communication in China," cited in note 8, which includes information about modes of conveyance.

57. Rudyard Kipling, *From Sea to Sea*, in *The Writings in Prose and Verse of Rudyard Kipling* (New York: Charles Scribner's Sons, 1897–1937), vol. 15, 339–341. Five local tour guides are listed in R. C. Hurley, *The Tourists' Guide to Canton, the West River and Macao*, 2nd ed. (Hong Kong: R. C. Hurley, 1903), 24. Three of them were from the A-Cum family, confusingly listed as A-Cum (senior), A-Cum, and A-Cum (junior).

58. Cooper, *Travels of a Pioneer of Commerce*, 15. Many Chinese brought up in missionary education were engaged in international trade in the treaty ports because they had connections with Westerners and knew foreign languages. Yen-P'ing Hao, *The Comprador in Nineteenth Century China: Bridge between East and West* (Cambridge, Mass.: Harvard University Press, 1970), 197–198.

59. Raphael Pumpelly, *My Reminiscences*, 2 vols. (New York: Henry Holt and Company, 1918), vol. 2, 469.

60. Joseph Dalton Hooker, *Himalayan Journals: Notes of a Naturalist in Bengal, the Sikkim and Nepal Himalayas, the Khasia Mountains, etc.*, 2 vols. (New Delhi: Today and Tomorrow's Printers & Publishers, 1969 [1854]), 105, 168–169, 219.

61. Robert Fortune, *A Residence among the Chinese: Inland, on the Coast, and at Sea. Being a Narrative of Scenes and Adventures during a Third Visit to China, from 1853 to 1856* (London: John Murray, 1857), 60, 97–98.

62. Collingwood, *Rambles of a Naturalist*, 71, 87.

63. Arthur Adams, *Travels of a Naturalist in Japan and Manchuria* (London: Hurst and Blackett, 1870), 73–75.

64. John Henry Leech, *Butterflies from China, Japan, and Corea*, vol. 1 (London: R. H. Porter, 1892–94), xxiv–xxvi.

65. William Gardener, "Robert Fortune: Methods of Collection," *Gardeners Chronicle Gardening Illustrated*, 9 November 1963, 331; idem, "Robert Fortune: His 'Favourite Chekiang,'" *Gardeners Chronicle Gardening Illustrated*, 12 October 1963, 270.

66. Fortune, *A Residence among the Chinese*, 60–62. Apparently herb-collectors were not rare in places where important herbs grew. When Antwerp Pratt finally reached the Tibetan border after a long journey, he found himself resting shoulder to shoulder with "nearly fifty" Chinese herb-collectors in a hut. Pratt, *To the Snows of Tibet through China*, 187–188.

67. Theophilus Sampson, *Botanical and Other Writings on China, 1867–1870* (Hamburg: C. Bell Verlag, 1984), 38.

68. Robert Swinhoe, "The Ornithology of Formosa, or Taiwan," *Ibis* 5 (1863): 207. On Chinese hunters, see Fortune, *A Journey to the Tea Countries*, 151–153; Jernigan, *Shooting in China*, 205–234, which describes in detail the weapons used by the Chinese hunters. A Chinese named Kum Ayen contributed a chapter on "Some Chinese Methods of Shooting and Trapping Game" to Wade, ed., *With Boat and Gun in the Yangtze Valley*, 178–185. He wrote, "A foreign sportsman is usually fully equipped with a fowling-piece with the latest improvements, the newest and best ammunition and a good dog. . . . On the other hand, the Chinaman takes the field clad in straw sandals and the poorest of clothes and uses a common, roughly-made gingal [the large Chinese matchlock], inferior native powder, and shot of unequal sizes. The foreigner shoots solely for pleasure, the native for a livelihood: possibly the sale of his bag goes to supporting parents, wife and family" (pp. 182–183).

69. Bretschneider, *Notes on Some Botanical Questions Connected with the Export Trade of China* (Peking, 1880), 5.

70. Kew: A. Henry Letters to H. B. Morse, ff. 7–9, 10–12.

71. Ibid.

72. C. B. Rickett and J. D. D. La Touche, "Additional Observations on the Birds of the Province of Fohkien," *Ibis*, sr. 7, 2 (1896): 489–490.

73. F. W. Styan, "Notes on the Ornithology of China," *Ibis*, sr. 6, 6 (1894): 329. It should be noted, however, that the French Consul Claude P. Dabry de Thirsant's native collector, a Catholic convert familiar with Chinese texts on animals, had done the feast—collecting fish, in this case—three decades before. Armand David, *Abbé David's Diary* (Cambridge, Mass.: Harvard University Press, 1949), 198.

74. Kew: Chinese and Japanese Letters, 151 (588–590), (601–602), (610) (620), (622), (623), (624).

75. Kew: Chinese and Japanese Letters, 151 (622).

76. Kew: A. Henry. Letters to H. B. Morse, ff. 34–35, 47–48, 119–120, 121, 124.

77. Bretschneider, *Notes on Some Botanical Questions*, 5.

78. Kew: A. Henry. Letters to H. B. Morse, ff. 10–12, 17–18. On Henry's botanical research in Taiwan, see Augustine Henry, "A List of Plants from Formosa, with Some Preliminary Remarks on the Geography, Nature of the Flora and Economic Botany of the Island." Published as supplement to *Transactions of the Asiatic Society of Japan* 24 (1896): 1–118.

79. Brian D. Morley, "Augustine Henry: His Botanical Activities in China, 1882–1890," *Glastra* 3 (1979): 21–81 (esp. 57, 75–77).

80. Henry to Sargent, 9 May 1899, in NBG: Augustine Henry Papers, file

"C. S. Sargent." The expenses were somewhat higher in the areas about the coastal treaty ports. The budget of Charles Ford's six-week expedition in Guangdong in 1882 listed 220 silver dollars for travel expenses, and the team consisted of only two or three people besides himself. Kew: Misc. Reports. 4.41. Hong Kong. Miscellaneous, ff. 53–57.

81. NHML: Z. Keeper's Archives. 1.29. Letters. 1886. January to June, No. 390.

82. Arthur de Carle Sowerby, *China's Natural History: A Guide to the Shanghai Museum* (Shanghai: Royal Asiatic Society, North China Branch, 1936), 3–4. Taxidermy was essential to the maintenance of a natural history museum. On taxidermy in the nineteenth century, see Paul L. Farber, "The Development of Taxidermy and the History of Ornithology," *Isis* 68 (1977): 550–566; P. A. Morris, "A Historical Review of Bird Taxidermy in Britain," *Archives of Natural History* 20 (1993), 241–255; Karen Wonders, "Bird Taxidermy and the Origin of the Habitat Diorama," in *Non-Verbal Communication in Science prior to 1900*, ed. Renato G. Mazzolini (Firenze, Italy: Leo S. Olschki, 1993), 411–447. F. W. Styan contributed a chapter on bird taxidermy to Wade, *With Boat and Gun in the Yangtze Valley*, 128–131. Although the educational impact of the Shanghai Museum was probably limited, it surely impressed the Chinese with its curious pictures and specimens. A Chinese visitor recorded his impressions of the museum in a poem, marveling at a huge snake, probably a python, preserved in a glass jar and reporting that "birds and animals are skinned, depicted, and recorded in drawings and texts, and the data and objects are, it is said, to be transmitted to the museums in the West." Gu Bingquan, ed., *Shanghai yangchang zhuzhici* (Shanghai: Shanghai shudian, 1996), 80. See also p. 99.

83. J. D. D. La Touche, "The Collection of Birds in the Shanghai Museum," *JNBC* 40 (1909): 69–107. Tang Wang-wang had long learned to draw sketches of bird's nests, note down observations, and perform other work in the field. Rickett kept one of his sketches of bird nests. C. B. Rickett, "Notes on the Birds of Fohkien Province, S. E. China," NHML: Tring. MSS. Rickett, ff. 180–181, 201–202.

84. Sowerby, *China's Natural History: A Guide to the Shanghai Museum*, 4.

85. My consideration of power, culture, and fieldwork has been influenced by a number of works in contemporary anthropology. For example, Akhil Gupta and James Ferguson, eds., *Culture, Power, Place: Exploration in Cultural Anthropology* (Durham: Duke University Press, 1997); idem, *Anthropological Locations: Boundaries and Grounds of a Field Science* (Berkeley: University of California Press, 1997); Sherry Ortner, "Thick Resistance: Death and the Cultural Construction of Agency in the Himalayan Mountaineering," *Representations* 59 (Summer 1997): 135–162; idem, "Resistance and the Problem of Ethnographic Refusal," in *The Historical Turn in the Human Sciences*, ed. Terrence J. McDonald (Ann Arbor: University of Michigan Press, 1996), 281–304.

86. The British (and the subjects of several other nations) did have the privilege of extraterritory, but this and their other diplomatic privileges, which sheltered them from the Chinese judicial system, had little direct impact on the actual working of the natural historical fieldwork in China unless it involved the law.

87. Cooper, *Travels of a Pioneer of Commerce*, 365–385.

88. Pratt, *To the Snows of Tibet through China*, 194–195.

89. Ibid., 5.

90. Ibid.

91. E. H. Wilson, *A Naturalist in Western China*, vol. 1, 86.

92. Ibid.

93. Henry to Sargent, 14 November 1899, NBG: Augustine Henry Papers, file "C. S. Sargent."

94. For example, Wade, ed., *With Boat and Gun in the Yangtze Valley*, 132–134. A review of Francis Groom, *The Sportsman's Diary for Shooting Trips in Northern China* (Shanghai, 1873) cites from the book many "hints" at how to proceed in these situations. *The North China Herald*, 20 November 1873. Unfortunately, I have not been able to find a copy of the book.

95. John K. Fairbank and Liu Kwang-Ching, eds., *The Cambridge History of China*, vol. 10, part II, 82–84.

96. Paul Cohen, *China and Christianity: The Missionary Movement and the Growth of Chinese Antiforeignism, 1860–1870* (Cambridge, Mass.: Harvard University Press, 1963); idem, "Christian Missions and Their Impact to 1900," in John K. Fairbank, ed., *The Cambridge History of China*, vol. 10, part I, 543–590; Zhonghua wenhua fuxing yundong tuixing weiyuanhui, ed., *Jiaoan yu fanxijiao*, vol. 4 in *Zhongguo jindai xiandai shi lunji* (Taipei: Taiwan shangwu, 1985); Su Ping, *Yaoyan yu jindai jiaoan* (Shanghai: Shanghai Yuandong chubanshe, 2001).

97. Joseph Esherick, *The Origins of the Boxer Uprising* (Berkeley: University of California Press, 1987). See also Paul A. Cohen, *History in Three Keys: The Boxers as Event, Experience, and Myth* (New York: Columbia University Press, 1997).

98. Thomas Stevens, *Around the World on a Penny-Farthing* (London: Arrow Books, 1991 [1888]), 383–385.

99. Wilson, *A Naturalist in Western China*, vol. 2, 3.

100. Pratt, *To the Snows of Tibet through China*, 49.

101. Ibid., 135, 151, 194, 222.

102. Royal Botanic Gardens, Edinburgh. George Forrest Papers. ARC/FOR/BAL/01–45. W. J. Embery to John Balfour, 17 August 1905; G. Forrest to Balfour, 29 August 1905; Forrest to Balfour, 1 October 1905.

103. Robert Swinhoe, "Notes on the Island of Formosa" (read before the BAAS, August 1863), 12. I used the bound volume of Swinhoe's papers on Formosa at the British Library. The paper was subsequently published in *Proceedings of the Royal Geographic Society* 8 (1864): 23–28.

104. Albert S. Bickmore, "Sketch of a Journey from Canton to Hankow through the Province of Kwangtung, Kwangsi, and Hunan, with Geological Notes," *JNCB*, n.s., 4 (1867): 9.

105. Michael Taussig discusses Indian *silleros* in his *Shamanism, Colonialism, and the Wild Man: A Study in Terror and Healing* (Chicago: University of Chicago Press, 1987), 287–335.

106. See, for example, the illustration "Superior British attitudes greatly annoyed the Chinese," in Welsh, *A History of Hong Kong*, between pp. 384–385; Clark Worswick and Jonathan Spence, *Imperial China: Photographs, 1850–1912* (London: Scolar Press, 1979), 82. See also ibid., 79, the "ethnographic/orientalist" picture of a Chinese mandarin carried in a sedan chair.

107. The contemporaries were highly conscious of the multiple meanings of taking the chair transport. For example, an American naval surgeon who visited China in 1856–1858 despised the "English" practice of using the sedan chair in Chinese cities. "On a slave plantation, or in any city of a southern State," he said,

"the most delicate and fragile lady would be ashamed to make a beast of burden of the negro slave, whilst it is not at all improbable that the two heavy old or new Englishmen now promenading upon the backs of these sweating Chinamen, are denouncing the horrors of American slavery." What about Americans in China? He patriotically announced that "[so] far as my own limited observation goes . . . gentlemen from the slave States of our South are more repugnant in China to making beasts of burden of human beings than are any other foreigners." After repudiating the possible reasons for "this painful indolence," including the horror of the sun, he concluded that "[these] chairs are a part of the moral and social system of the Chinese, and hence there are reasons for their use, which foreigners have not." William Maxwell Wood, *Fankwei; or, the San Jacinto in the Seas of India, China and Japan* (New York: Harper & Brothers, 1859), 365–371. The Chinese officials had interpreted the privilege of riding a sedan chair in their own cultural terms. In the times of Old Canton, they forbade Western traders to use the mode of transport as a means to assert, so to speak, their view of world order, and the Westerners protested repeatedly against this rule, which they perceived as humiliating. H. B. Morse, *The East India Company Trading to China*, vol. 4, 81, 234, 236, 244, 298, 348–349.

108. *Report by Mr. Baber on the Route followed by Mr. Grosvenor's Mission between Tali-Fu and Momein.* Parliamentary Papers. China. No. 3 (1878), 5, 21. Baber would receive recognition from the Royal Geographic Society for this and other explorations in China.

109. *Report by Mr. Hosie of a Journey through Central Ssu-Ch'uan in June and July, 1884.* Parliamentary Papers. China. No. 2 (1885), 13.

110. George Orwell, *A Collection of Essays* (San Diego: Harcourt Brace & Company, 1946), 156. Scholars have argued the importance of imperial/colonial context to the formation of "Englishness." See, for example, Manu Goswami, "'Englishness' on the Imperial Circuit: Mutiny Tour in Colonial South Asia," *Journal of Historical Sociology* 9 (1996): 54–84; Kathryn Tidrick, *Empire and the English Character* (London: I. B. Tauris, 1990); Ian Baucom, *Out of Place: Englishness, Empire, and the Locations of Identity* (Princeton: Princeton University Press, 1999); Simon Gikandi, *Maps of Englishness: Writing Identity in the Culture of Colonialism* (New York: Columbia University Press, 1996).

111. Swinhoe, of course, did not have Orwell's Edwardian anti-imperialist sentiment. Robert Swinhoe, "Catalogue of the Mammals of China (South of the River Yangtze) and the Island of Formosa," *Proc. Zool. Soc.* (1870): 626–627. See also "The Tiger in Amoy," *The Zoologist* 19 (1861): 7701–7702. A sportsman said of the villagers near Amoy that "[if] a man has been killed [by the tiger] they welcome the foreign sportsman most cordially, and chin-chin him as the saviour of the people, but if the tiger has not been known to take human life they are disposed to propitiate him, and condemn the folly of enraging him." R. H. Bruce, "Sport in Amoy," *China Review* 21 (1894–1897): 718–719.

112. Along these lines, Bernard S. Cohen discusses colonialism and dress codes in the essay "Cloth, Clothes, and Colonialism: India in the Nineteenth Century," collected in his *Colonialism and Its Forms of Knowledge: The British in India* (Princeton: Princeton University Press, 1996), 106–162.

113. *Report by Mr. Hosie of a Journey through the Provinces of Ssu-ch'uan, Yunnan, and Kuei Chou*, 17.

114. James, *The Long White Mountain*, 230.

115. Cooper, *Travels of a Pioneer of Commerce*, 9.

116. Fortune, *Three Years' Wanderings in China*, 197–224; idem, *A Journey to the Tea Countries of China*, 272–287.

117. For example, James Secord, "Nature's Fancy: Charles Darwin and the Breeding of Pigeons," *Isis* 72 (1981): 163–186; "Darwin and the Breeders: A Social History," in David Kohn, ed., *The Darwinian Heritage* (Princeton: Princeton University Press, 1985), 519–542.

118. Swinhoe, "The Natural History of Hainan," 13. I used the reprinted copy in the bound volume titled "The Natural History of Hainan" at the British Library. The paper first appeared in the magazine *Field* in 1870.

119. Robert Swinhoe, "The Ornithology of Formosa, or Taiwan," *Ibis* 5 (1863): 198–219, 250–311, 377–435. On pp. 394–395.

120. Ibid., 208–209; idem, "On the Mammals of the Island of Formosa, (China)," *Proc. Zool. Soc.* (1862): 347–365, esp. 348–350; idem, "The Natural History of Hainan," 13–15.

121. D. E. Bufford and S. A. Spongberg, "Eastern Asian-Eastern North American Phytogeographical Relationships—a History from the Time of Linnaeus to the Twentieth Century," *Annals of Missouri Botanical Gardens* 70 (1983): 423–439; Philip J. Pauly, *Biologists and the Promise of American Life: From Meriwether Lewis to Alfred Kinsey* (Princeton: Princeton University Press), 15–43.

122. When purchasing tea seeds in Zhejiang, for example, Fortune needed the help of a Buddhist monk, who was "an excellent judge of the seed." Fortune, *A Residence among the Chinese*, 143. Similar incidents were numerous.

123. Alfred Newton, "Abstract of Mr. J. Wolley's Researches in Iceland respecting the Gare-fowl or Great Auk (*Alca impennis*, Linn.)," *Ibis* 3 (1861): 374–399.

124. Swinhoe, "The Ornithology of Formosa, or Taiwan," 207.

125. For example, Arthur Adams, William Balfour Baikie, and Charles Barron, *A Manual of Natural History for the Use of Travellers* (London: John van Voorst, 1854), 632–637, 646–647, 656–662, 676–681.

126. Armand David, *Abbé David's Dairy*, 23–26, 38, 165–166, 198, 206, 225.

127. Ibid., 38.

128. For example, ibid., 276–283.

129. Swinhoe, "The Ornithology of Formosa or Taiwan," 275.

130. Alfred Russel Wallace, *The Malay Archipelago, the Land of the Orang-Utan and the Bird of Paradise: A Narrative of Travel with Studies of Man and Nature* (New York: Dover Publications, 1962 [1890]), 29–30.

131. For example, Fortune, *Three Years' Wanderings in China*, 135.

132. Sampson, *Botanical and Other Writings on China*, 9.

133. T. L. Bullock, "Canton and Peking Plants," *China Review* 17 (July 1888–June 1889): 307–308.

134. Sampson, *Botanical and Other Writings on China*, 17.

135. Charles Ford to Thiselton-Dyer, 9 April 1891. Kew: Misc. Reports. 4.4. China: Foods, Medicines, & Woods, 1869–1914, ff. 545–550. Ford remarked: "From my somewhat extended experience with Chinese in [China and Hong Kong] I know how little reliance is to be placed on information supplied by the ordinary Chinaman in regard to plants. I would not withhold due acknowledgements of the usefulness of the natives in helping us to get at true information, but their aid should only be regarded as collateral, the investigator himself should shift & verify everything important."

136. On the role of orientalism in the formation of a Western identity, see, for example, James G. Carrier, *Occidentalism: Images of the West* (Oxford: Clarendon Press, 1995).

137. E. H. Parker, "Canton Plants," *China Review* 15 (1886–87): 104–119. The piece was originally published in the *China Mail*, a Hong Kong newspaper, in 1878.

138. Ibid.

139. Robert Swinhoe, "Notes on Chinese Mammalia Observed near Ningpo," *Proc. Zool. Soc.* (1872): 813; idem, "The Small Chinese Lark," *JNCB*, sr. 1 (December 1895): 291–292. Henry Hance also had a similar experience. After asking a druggist in Canton about some drugs, Hance was surprised to find that "strange as it may seem, it would really look as if the Chinese did not distinguish the two [plants]!" Hance to Daniel Hanbury. 27 February 1867, RPS: Hanbury Papers. P313 [3].

140. Sampson once complained, "There is not a common weed that is not utilized by the herbalist as 'medicine' and that is all that a Chinaman knows or cares to know about the vegetable world. Chinese gentlemen in the interior always in towns. . . . Such a thing as a taste for botany is utterly unknown to the Chinese mind; you may instill some idea of botanical requirements into the minds of coolies who work for you, but it is hopeless to expect anything whatever from . . . any Chinese gentlemen." Sampson to Ford, Kew: Misc. Reports. 4.41. Hong Kong. Miscellaneous, f. 50.

141. Alexander Hosie, *Report by Consul-General Hosie on the Province of Ssuchúan*. Parliamentary Papers. China. No. 5 (1904), 52.

142. Parker, "Canton Plants," 116.

143. La Touche to Oldfield Thomas, 2 August 1898, NHML: Curator of Mammals. Correspondence. DF232/5, ff. 514–515.

144. Oldfield Thomas, "On Mammals Collected by Mr. J. D. La Touche at Kuatun, N. W. Fokien, China," *Proc. Zool. Soc.* (1898): 770.

145. For example, Fortune, *A Residence among the Chinese*, 98–99.

146. C. B. Rickett, "Notes on the Birds of Fohkien Province, S. E. China," NHML: Tring. MSS. Rickett, ff. 234, 246.

147. He did, however, jokingly offer a ten-dollar award for a specimen of the one-legged bird. NHML: Tring. MSS. Rickett, f. 246.

148. David, *Abbé David's Diary*, 23–24.

149. Shapin, *A Social History of Truth*, chs. 5 and 6.

150. For example, Swinhoe, "On the Mammals of the Island of Formosa," 347–365 (on pp. 360, 361); idem, "The Ornithology of Formosa, or Taiwan," 401.

151. Peter Hopkirk, *The Great Game: The Struggle for Empire in Central Asia* (New York: Kodansha International, 1992). Explorers working for the British, including the famous Pundits, frequently ventured into China from India, central Asia, and Indo-China. See Gabriel Finkelstein, "'Conquerors of the Künlün'? The Schlagintweit Mission to High Asia, 1854–57," *History of Science* 38 (2000): 179–218; John MacGregor, *Tibet: A Chronicle of Exploration* (London: Routledge & Kegan, 1970); Derek Waller, *The Pundits: British Exploration of Tibet and Central Asia* (Lexington: University Press of Kentucky, 1990). On the French activities, see Numa Broc, "Les voyageurs français"; idem, "Les explorateurs français"; Fournier, *Voyages et découvertes scientifiques*. The French also tried to find routes into China from Indo-China. For example, Milton Osborne, *River to China: The Mekong River Expedition, 1866–1873* (London: George Allen and Unwin, 1975). See also Victor

T. King, ed., *Explorers of South-East Asia: Six Lives* (New York: Oxford University Press, 1995). Bretschneider, *History* includes all the major Russian explorers in China.

152. On landmarks and exploration, see, for example, D. Graham Burnett, *Masters of All They Surveyed: Exploration, Geography, and a British El Dorado* (Chicago: University of Chicago Press, 2000), 119–198.

153. Jeremy Black, *The Grand Tour in the Eighteenth Century* (Stroud: Sutton Publishing, 1992); Christopher Hibbert, *The Grand Tour* (London: Thames Methuen, 1987); John Pemble, *The Mediterranean Passion: Victorians and Edwardians in the South* (Oxford: Oxford University Press, 1988); James Buzard, *The Beaten Track: European Tourism, Literature, and the Ways to "Culture," 1800–1918* (Oxford: Oxford University Press, 1993).

154. William Langdon, *Ten Thousand Things relating to China and the Chinese* (London: To be had only at the collection, Hyde Park Corner, 1842). The high-minded Carlyles did not enjoy the exhibition, partly because "it was crowded to a degree which made it impossible to see the *tools* and other particulars which alone deserved the notice of Literary gentlemen." Clyde de Ryals and Kenneth Fielding, eds., *The Collected Letters of Thomas and Jane Welsh Carlyle*, vol. 16 (Durham: Duke University Press, 1990–), p. 19. On both the Chinese exhibition and the junk, see Richard Altick, *The Shows of London* (Cambridge, Mass.: Harvard University Press, 1978), 292–297. James R. Ryan, *Picturing Empire: Photography and the Visualization of the British Empire* (Chicago: University of Chicago Press, 1997), ch. 5, includes a discussion of photographing and travel in China. On the issue of space, time, and representation of the Other, see Johannes Fabian, *Time and the Other: How Anthropology Makes Its Object* (New York: Columbia University Press, 1983).

155. Michael Dettelbach, "The Face of Nature: Precise Measurement, Mapping, and Sensibility in the Work of Alexander von Humboldt," *Studies in the History and Philosophy of Biological and Biomedical Sciences* 30C (1999): 473–504.

156. The literature on travel writing and imperialism is extensive. I have benefited especially from the following works: Peter Bishop, *The Myth of Shangri-La: Tibet, Travel Writing and the Western Creation of Sacred Landscape* (Berkeley: University of California Press, 1989); Sybille Fritzche, "Narrating China"; Billie Melman, *Women's Orient: English Women and the Middle East, 1718–1918* (Ann Arbor: University of Michigan Press, 1992); Ali Behad, *Belated Travelers: Orientalism in the Age of Colonial Dissolution* (Durham: Duke University Press, 1994); Dennis Porter, *Haunted Journey: Desire and Transgression in European Travel Writing* (Princeton: Princeton University Press, 1991); Carter, *The Road to Botany Bay*; Pratt, *Imperial Eyes*; David Spurr, *The Rhetoric of Empire: Colonial Discourses in Journalism, Travel Writing and Imperial Administration* (Durham: Duke University Press, 1993); Tzvetan Todorov, *On Human Diversity: Nationalism, Racism, and Exoticism in French Thought* (Cambridge, Mass.: Harvard University Press, 1993), ch. 4; Michael T. Bravo, "Ethnological Encounters," in Jardine, Secord, and Spray, eds., *Cultures of Natural History*, 338–357.

157. Peter Perdue, "Boundaries, Maps, and Movement: Chinese, Russian, and Mongolian Empires in Early Modern Central Eurasia," *International History Review* 20 (1998): 263–286; Theodore N. Foss, "A Western Interpretation of China: Jesuit Cartography," in *East Meets West: The Jesuits in China, 1582–1773*, ed. Charles E. Ronan and Bonnie B. C. Oh (Chicago: Loyola University, 1988), 209–251;

Michele Pirazzoli-T'Serstevens, *Gravures des conquêtes de l'empereur de Chine Kien-Long au Musée Guimet* (Paris: Musée Guimet, 1969). See also Joanna Waley-Cohen, *The Sextants of Beijing: Global Currents in Chinese History* (New York: W. W. Norton, 1999), ch. 3.

158. Waley-Cohen, *The Sextants of Beijing* and her "China and Western Technology in the Late Eighteenth Century," *American Historical Review* 98 (1993): 1525–1544.

159. For example, Fortune, *Three Years' Wandering in China*, 134; idem, *A Journey to the Tea Countries*, 231.

160. One feels that the fishermen in Bruno Latour's discussion of "centers of circulation" deserve more sophisticated discussion than they receive. Bruno Latour, *Science in Action* (Cambridge, Mass.: Harvard University Press, 1987), 215–223.

Epilogue

1. Alfred Russel Wallace, *My Life, a Record of Events and Opinions*, vol. 2 (London: Chapman & Hall, 1905), 60–61.

2. Georges Métailié, "Comparative Study of the Introduction of Modern Botany in Japan and China," *Historia Scientiarum* 11 (2002): 205–217.

3. For example, E. H. M. Cox, *Plant-Hunting in China*, chs. 5 and 6; Reginald Farrer, *The Rainbow Bridge* (London: Cadogan Books, 1986 [1926]); E. H. Wilson, *A Naturalist in Western China* (London: Cadogan Books, 1986 [1913]); idem, *China: Mother of Gardens* (New York: Benjamin Blom, 1971 [1929]); Roy Briggs, *"Chinese" Wilson: A Life of Ernest H Wilson, 1876–1930* (London: HMSO, 1993); J. Macqueen Cowan, ed., *The Journeys and Plant Introduction of George Forrest* (Oxford: Oxford University Press, 1952); Isabel S. Cunningham, *Frank N. Meyer: Plant Hunter in Asia* (Ames, Iowa: Iowa State University Press, 1984); Charles Lyte, *Frank Kingdon-Ward: The Last of the Great Plant Hunters* (London: John Murray, 1989); Arthur de Carle Sowerby, *Sport and Science on the Sino-Mongolian Frontier* (London: Andrew Melrose, 1918); idem, *Fur and Feather in North China* (Tientsin [Tianjing]: Tientsin Press, 1914); idem, *A Naturalist's Note-Book in China* (Shanghai: North-China Daily News & Herald, 1925); idem, *The Naturalist in Manchuria*, 3 vols. (Tientsin [Tianjing]: Tientsin Press, 1922–23); R. R. Sowerby, *Sowerby of China* (Kendal: Titus Wilson and Son, Ltd., 1956). Bernard Read and his Chinese colleagues did much research on traditional Chinese herbals in the 1920s and 1930s, and many of their articles first appeared in *Peking Natural History Bulletin*. For example, Bernard Read and Li Yü-Thien, "Chinese Materia Medica; I–V. Animal Drugs," 5 (1931): 37–80; 6 (1932): 1–120. Not until the late 1930s were the activities of British naturalists in China terminated by World War II and finally by the Communist government, which came into power in 1949.

4. Lynn K. Nyhart, "Natural History and the 'New' Biology," in Jardine, Secord, and Spray, eds., *Cultures of Natural History*, 426–443.

5. Some of these issues are considered in my "Nature and Nation in Chinese Political Thought: The National Essence Circle in Early Twentieth-Century China," in Lorraine Daston and Fernando Vidal, eds., *The Moral Authority of Nature* (Chicago: University of Chicago Press, 2004).

Index

Abel, Clarke, 19, 35, 37
Adams, Arthur, 125, 133
Aesthetics, 13, 38
Africa, 24, 129
Agassiz, Louis, 145
Agency, 4, 86–87, 138, 157
Alabaster, Chaloner, 74, 199n69
Alcock, Rutherford, 82–83
Allen, Young John, 158
Alligators, 115–118, 120
Americans (U.S. citizens), 15, 73, 79, 140, 226n107
Amherst Embassy, 18, 19
Amoy (Xiamen), 18, 68, 73
Anatomical investigation, 53, 93
Andrews, Roy Chapman, 126
Animals, 14, 22, 77, 83; in Chinese literature, 100, 101, 105, 106, 109–110, 116–117; classification of, 111; domesticated, 120; drawings of, 49; folk knowledge concerning, 146; geographical distribution of, 102–103; shipboard, 37; sold in marketplace, 27, 28–29, 180n89
Archaeology, 159
Art. *See* Natural history illustration
Audubon, John, 42, 117
Australia, 24, 129

Baber, E. C., 142
Banks, Joseph, 14, 18, 20, 155; gardener-collectors and, 24; horticulture and, 62; natural history drawings and, 46–47, 50; scientific commonweal ideal and, 35; transporting of plants and, 37
Bassett-Smith, P. W., 125
Bates, Henry Walter, 124
Beale, Thomas, 23, 25, 26, 34, 44–45; aviary of, 28, 45, 50; biographical notes on, 163; death of, 45
Beijing, 24, 80, 129
Bencao gangmu (Li Shizhen), 50, 94, 102, 107, 109; on Chinese alligator, 117; on domesticated animals, 120; known in the West, 95; scope and reputation of, 105–106
Bennett, George, 45, 50
Bentham, George, 71, 72, 105
Betel palm, illustration of, 53–54
Biogeography, 102–103, 144–145
Bird, Isabella, 129
Birds, 27, 89, 136; Beale's aviary, 28, 45, 50; in Chinese literature, 106; classification of, 144, 148; drawings of, 42, 50–51, 52; folk knowledge concerning, 149, 150; hunting of, 129–130
Blake, John Bradby, 46, 57
Borderlands, 3
Botanical Garden of St. Petersburg, 128
Botanical Magazine, 51
Botanicon sinicum (Bretschneider), 111

Botany, 67, 68, 72, 78, 91, 100; botanical geography, 102–103, 111; Chinese view of, 105; comparison of Western and Chinese systems, 106–109; distribution of trees, 145; economic, 62, 82, 86, 90, 99, 111; exploration and, 125; multivolume work on Chinese plants, 155; pharmacy and, 80; scientific journals devoted to, 99. *See also* Flowers; Plants

Boundaries, 3

Bowra, Edward C., 74, 78, 111–112

Bowring, John C., 81, 197n50

Boxer Uprising (1900), 140

Boym, Michael, 95

Bretschneider, Emil, 69, 78–79, 99, 111, 119; Chinese language and, 112; Chinese natural history works and, 107; collection of specimens and, 135, 136; on geographical botany, 103; plant names and, 117; scientific networks of, 100; use of gazetteer, 110

Bridgman, E. C., 97

Britain: botanical collectors sent around world, 24; diplomatic delegations to China, 18; horticultural societies in, 19–20; maritime power of, 3, 14, 38; natural history in, 2–3; professionalization of science in, 91; rivalry with other European empires, 151; social hierarchy of English society, 24–25; trade with China, 6

British Association for the Advancement of Science, 56

British Consular Service, 63, 64, 68, 69, 72; Chinese language used in, 96; commerce and, 83; dangers of travel and, 139; employees of, 73; naturalists in, 74; politics and, 75; scientific networks and, 75–76; tea manufacture and, 84

British Museum, 114, 120, 137, 155

British naturalists, 39, 74, 89; agency of indigenous people and, 157; Chinese artists and, 54–57; Chinese "associates" and employees of, 2, 40, 78, 150–151, 153; Chinese authorities and, 138; Chinese language and, 68, 98, 102; Chinese scholars and, 94; Darwinism and, 198n57; empire and, 63–64, 64, 90; exploration of Chinese interior, 7, 110; history of involvement in China, 5–6; Hong merchants and, 34; modes of collecting, 131–137; relations with Chinese population, 1–2, 85–87, 137–143, 147–148, 157; salaries paid to, 127; scientific networks and, 72, 75–80; as sinologists, 97–99; as traders, 17–23; visual authority of illustrations and, 54–55

Buffon, Georges-Louis, 11

Buitenzorg Botanical Gardens (Java), 79

Burma, 139, 142, 151

Cambridge University, 21, 45

Candolle, Alphonse de, 103, 119

Canton (Guangzhou), 1, 6, 15–16, 38–39; artisans in, 47; as contact zone, 16–17; culture of travel and, 129; garden nurseries of, 14; journal published in, 97; markets of, 26–31, 56, 180n89; suburbs of, 33, 47; tour guides in, 132; Western trade confined to, 14–15

Canton Factories, 15, 16, 17, 27, 33

Canton Factory (of the English East India Company), 19, 20, 22, 23; disbandment of, 22; gardener-collectors and, 24, 25; Reeves and, 43

Cantor, Theodore, 57

Capital, natural history and, 12

Carlyle, Thomas and Jane, 152, 229n154

Carpenter Square, 17

Castiglione, Giuseppe, 47

Catholic missionaries, 14, 80, 135, 198n57. *See also* Jesuit missionaries

Central Asia, 103–104, 114

Chalmers, John, 81

Chambers, William, 13

China: British trade with, 3; contacts with other countries, 103–104; in global context, 5; literature and myth of, 100–101; natural history of, 2, 90, 95; population of, 123, 153; religions of, 113; Republican, 160; scientific explorations in, 125–126; "surveying" of, 83; traditions of "natural history" in, 104–111; travel literature about, 152; travel restrictions on foreigners, 101; Western culture and, 3–4; Western fascination with, 13–14. *See also* Qing Empire; *specific cities and provinces*

China illustrata (Kircher), 95

China Inland Mission, 80–81

China Review (journal), 97

China trade, 11, 12, 13, 23, 82

Chinese artists, Reeves's (Akam, Akew, Akut, and Asung), 49

Chinese Classics, The (Legge), 94

Chinese language, 18, 22, 73, 94–95;

Index 233

learned by Western sinologists, 112; science taught in, 113; studied in Britain, 96; studied in Europe, 95
Chinese Maritime Customs, 63, 73, 74, 83, 102; British management of, 87; Chinese language used by, 96; *materia medica* issued by, 79; relation to British scientific organizations, 74–75; scientific networks and, 75–76
Chinese nurserymen: Aching, 30, 85; Old Samay, 30
Chinese people, 4–5, 157; artisans and artists, 47–49, 54, 192n84; British scientific imperialism and, 83–89; Christian converts, 135, 140, 150; employed by British naturalists, 2, 25, 137; gardeners and herbalists, 148, 149; guides and informants, 124, 131–132, 137, 138–139, 145, 149–150, 153; orientalist discourse and, 88; participation in scientific enterprise, 93–94; scholars, 158; wages paid to, 136–137; Western distrust of, 147–148
Chinese Recorder (journal), 97
Chinnery, George, 48
Chinoiserie, 13
Chowqua, 33
Christianity, 113
Civil history, 100
Civilization, 22, 23, 123
Classification. *See* Taxonomy
Coastal surveys, 125
Collingwood, Cuthbert, 125, 132, 163
Colonial Office (London), 65, 66
Concession leases, 63
Consequa, 33, 34, 182n116
Contact zone, 3
Cooper, Thomas T., 132, 138, 143
Crane, symbolism of, 54
Cultural encounter, 137–143
Cunningham, James, 46

Darwin, Charles, 45, 71, 103, 119–120, 144
Darwinism, 71–72, 145
David, Armand, 80, 99, 105, 123, 131; Chinese collector employed by, 137; folk knowledge and, 146, 150; on Protestant missionaries, 202n100; travels in Mongolia and Tibet, 132
David's deer, 82–83
Davis, John, 88
De Guignes, Joseph, 95
Delavay, Jean Marie, 80, 131

Descent of Man, The (Darwin), 120
Du Halde, Jean-Baptiste, 13, 95
Duncan, Alexander, 20, 35, 36, 46, 47, 163
Duncan, John, 20
Dutch East India Company, 12, 173n11
Dutch East Indies, 79–80

Edgerton, Samuel, 42
Edinburgh, Royal Botanic Gardens, 140
Ehret, Georg Dionysius, 42
Elements of Botany (Lindley), 105
English East India Company, 12, 15, 19, 34; Banks protégés in, 20; end of trade monopoly, 22; gardener-collectors and, 24, 26; Reeves and, 43; tea manufacture and, 84
Enlightenment, 95
Entomology, 28, 133
Erya (ancient Chinese dictionary), 108
Ethnography, 100
Ethnoscience, 99
Eurocentrism, 93, 115
Export painting, 47–49, 50
Extraterritoriality, 63, 224n86

Faber, Ernst, 81, 108, 117
Facts, 87–88
Fairbank, John K., 5
Fangzhi (gazetteer), 88, 109–110
Farrer, John Reginald, 159
Fa-tee nurseries, 14, 25, 27, 44, 179n80; Chinese New Year at, 30; variety of flowers cultivated, 29
Fauna Japonica (Siebold), 105
Fauvel, Albert A., 108, 115–118
Fieldwork, 6, 27, 81, 93, 137, 159; folk knowledge and, 143–151; naturalists' entourage in, 132; scientific imperialism and, 138; use of native guides and collectors, 83. *See also* Natural history
Fish, drawings of, 55–57, 192n75
Fitch, W. H., 55
Flora Honkongensis (Bentham), 105
Flowers, 18, 29, 35; British fascination with Chinese flowers, 14; drawings of, 46, 50–51, 52. *See also* Botany; Plants
Folk knowledge, 143–151
Forbes, Francis B., 69, 81
Ford, Charles, 1, 65–68, 85, 90, 102, 163
Forrest, George, 62, 127, 140, 159, 163
Fortune, Robert, 57, 62, 146; biographical notes on, 163; British-Chinese mistrust and, 85; in Chinese costume, 131; on

Fortune, Robert *(continued)*
  diversity of Chinese population, 153; folk knowledge and, 149; seed company cofounded by, 156; tea manufacture and, 84, 143–144, 145; travels and travelogues of, 86, 123, 126, 127–128, 152
Fossils, 11, 111
France, 17–18, 39, 198n57; Chinese language studies in, 95; French missionaries, 73, 80, 123, 127, 131, 140; in Indo-China (Tonkin), 85, 151; joint military operation with Britain, 61, 188n42
Franchet, Adrien, 105
Fraser, M. F. A., 84
Fryer, John, 158
Fujian Province, 149

Gardens and gardening, 13–14, 18, 105, 184n141; Chinese, 29, 184n141; gardener-collectors, 23–26
Genera plantarum (Bentham and Hooker), 71
Genre painting (Chinese), 47, 49
Geography, 3, 4, 109–110
Geology and geologists, 81, 82, 96, 125–126, 145, 159
German Consular Service, 106
Germans, 123, 140
Gould, John, 55
Gray, Asa, 106, 145
Gregory, William, 77
Guangdong Province, 33, 134
Guang qunfang pu ("Enlarged Botanical Thesaurus"), 109, 115
Guangxi Province, 84, 140, 141
Guizhou Province, 151
Guizhou tongzhi (Gazetteer of Guizhou), 110
Gunboat diplomacy, 63, 143

Hainan (island), 75, 79, 99, 110, 145, 155
Hanbury, Daniel, 79, 102
Hanbury, Thomas, 79
Hance, Henry Fletcher, 1, 57, 67–72, 76, 128, 148; biographical notes on, 163–164; on Darwinism, 71–72; herbarium of, 67, 69, 155; journal articles by, 99; scientific networks of, 77, 79–80, 100, 196n42; texts and, 111
Hancock, William, 69, 74, 164
Hardwicke, Thomas, 56
Hart, Robert, 73, 75, 87, 128
Henry, Augustine, 74, 75, 76, 109; biographical notes on, 164; collection of specimens and, 135; field trips of, 131; interest in Chinese medicine, 102; scientific networks of, 100; texts and, 117; travels on Yangzi River, 140
Henry, B. C., 81
Henslow, John, 21, 45, 85
Herbaria, 67, 69, 100, 155
Heude, Pierre Marie, 80, 99, 131, 198n57, 210n47
Ho Kai, 102
Holt, H. F. W., 74
Honam, gardens on, 33, 34
Hong Kong, 62, 81; foresting of, 65–66; journals published in, 97; Opium War and, 61; tour guides in, 132; Western residents of, 202n105
Hong Kong Botanic Gardens, 1, 63, 64–68, 83, 90, 195n28
Hong merchants, 15, 25, 31–35
Hooker, Joseph, 65, 68, 74, 83, 90, 91
Hooker, William, 14, 21, 68, 72, 128
Horticultural Society of London (Royal), 14, 22, 44, 125; gardener-collectors employed by, 24, 25, 27; natural history drawings and, 45; Reeves as member of, 44; scientific exploration sponsored by, 126, 127; transporting of plants and, 38
Horticulture, 2, 62, 156; British, 14, 30; Chinese, 18, 25; global maritime trade and, 38; Hong merchants and, 34, 35
Hosie, Alexander, 142, 149, 164
Howqua II, 32–33
Huatian ("Flower Fields") village, 29
Hubei Province, 62, 140
Huc, Évariste, 123, 132
Humanism, 100, 119
Humboldtian science, 102, 152
Hume, Abraham, 19
Hunan Province, 140
Hunting, 7, 82, 125, 139, 159; Chinese hunters and weapons, 223n68; folk knowledge and, 149; imperialism and, 220n44; Victorian culture and, 128–130
Hutcheson, Francis, 13
Hybridization, 3, 4, 47, 151

Ichthyology, 56–57
Impact-response model, 5
Imperialism, 17, 121, 157; botanical empire, 63; "informal empire," 62–63, 64, 72–75, 90; "muscular imperialism," 74, 141; planned partition of China, 151;

Index

resistance to, 86, 87, 138; symbolic ideology of, 141; travel accounts and, 124. *See also* Scientific imperialism
Index Florae Sinensis, 69
India, 41, 56, 132; cultural contacts with China, 103, 104; natural history illustration in, 191n67; plants transmitted to China, 110, 117; tea production in, 82, 84; tiger hunting in, 129
Indo-China, 85, 120, 151
Insects, 28, 130; collected by children, 132, 133; drawings of, 42, 50–51, 52, 190n58
Irish Protestants, 73

James, Henry E. M., 129
Japan, 77, 104, 126, 128
Jardine, William, 52
Jesuit missionaries, 2, 13, 17, 18, 80; artists, 47; botanical names and, 46; Chinese language and, 94; library at Xujiahui, 117; mapping of Qing territories and, 153; technical journal articles by, 99; writings on Chinese medicine, 95. *See also* Missionaries
Jiuhuang bencao (Chinese herbal), 108

Kames, Lord, 13
Kerr, William, 25, 34, 37, 47, 127, 164
Kew, Royal Botanic Gardens, 13, 14, 20, 21, 75, 91; botanical empire of, 62, 64–65; collection of Chinese plants, 155; gardener-collectors employed by, 24, 25; Hong Kong Botanic Gardens and, 66; primary objective of, 195n28; Protestant missionaries and, 81; reliance on Chinese knowledge, 84; restrictions on collectors, 219n27; scientific exploration sponsored by, 126–128, 136; transporting of plants and, 37, 38
Kingdon-Ward, Francis, 159
Kingsmill, Thomas, 81
Kipling, Rudyard, 132
Kircher, Athanasius, 11, 13, 94–95
Knowledge, 54–57, 119–121; empire and, 89–90; folk knowledge, 143–151; "objective" facts and, 64, 87–89
Kopsch, Henry, 77

Lamqua, 48–49
Language, 45–46, 93; cultural exchange and, 104; origin of, 95; pidgin, 16, 31, 35; translations, 158; Western languages, 113, 120, 132. *See also* Chinese language
Laos, 136

Latin, scientific, 41, 89, 107
La Touche, John David Digues, 75, 82, 130, 131, 137; biographical notes on, 164; collection of specimens and, 135; folk knowledge and, 149
Le Comte, Louis, 13
Legge, James, 94, 98
Leibniz, Gottfried, 95
"Lesser galangal," 79
Li Dou, 33
Lindley, John, 55, 105
Linnaeus, Carl, 56, 95, 107, 143
Linnean Society, 44
Linnean taxonomy, 89, 111, 148
Li Shizhen, 105
Litchi fruit, 15, 18, 21, 99
Literary Club, 20
Livingstone, David, 124
Livingstone, John, 22, 23, 25, 44, 164
Li Yuan, 94
Loddiges and Sons (nursery firm), 21

Macao, 15, 22, 23; British residents of, 44, 45; Chinese gardens of, 26; fish from markets of, 56; naturalists' excursions to, 27
Macartney, Lord, 20
Macartney Embassy, 18, 47
Main, James, 26, 30, 34, 46, 57, 164
Mammals, 52, 78, 82–83, 130, 136
Manchuria, 61, 123, 129, 142, 151, 155
Manson, Patrick, 73
Mapping, 89, 124, 152–153
Margary, Augustus, 139
Maries, Charles, 126, 140, 164
Maritime power, 3, 12, 14, 38
Martin, William J. P., 158
Martini, Martinus, 13
Masculinity, ideology of, 124
Materia medica, 78, 79, 81, 99, 101–102, 145
Maximowicz, Carl, 128
Mayers, William Frederick, 98, 114, 164
McGowan, D. J., 81
Meadows, Thomas T., 88
Medhurst, Walter, Sr., 81
Medicine, 22, 78, 79, 95, 101–102
Melbourne Botanic Gardens, 66
Merchants, 3, 15–16, 75, 156; as assistants to naturalists, 78; expansion of British power and, 62; Western (British), 81–82, 96
Merian, Maria Sibylla, 42
Meyer, Frank, 159
Minerals, 100, 105, 110

Ming Empire, 29, 32, 105
Minqua, 33
Missionaries, 3, 22, 43–44, 75, 129, 156; attacked by mobs, 140; Catholic, 14, 80, 135, 198n57; Chinese language and, 96; Darwinism and, 71; expansion of British power and, 62; French, 73, 80, 123, 127, 131; natural history contributions of, 80–83; Protestant, 63, 73, 80–81, 97; schools run by, 158; view of Chinese "heathenism," 113. *See also* Jesuit missionaries
Möllendorff, Otto F. von, 106, 108, 110
Monboddo, Lord, 95
Mongolia, 123, 132, 159
Morrison, Robert, 21, 22, 44, 96
Morse, Hosea, 74
Mowqua II, 32
Mueller, Ferdinand von, 66
Munqua, 34
Museums, 21, 22–23, 80, 93, 100, 144

Naming: imperialism and, 115, 152; as a reward, 77; sinology and, 98
Natural history: attractions of, 122; Chinese interest in Western science, 158–160; Chinese traditions of, 104–111, 113–114; folk knowledge and, 151; historical development of, 11–12, 155–156; hunting and, 128–130; knowledge translation and, 119–121; scope of, 100–104; sinology and, 92–94, 94–96. *See also* Fieldwork
Natural history illustration, 6, 41–43, 49–52; in Chinese literature, 108–109; communication and, 45–46; as site of cultural encounter, 52–54
Natural History Museum of London, 55–56
Natural History Museum of Shanghai, 82, 115, 126, 130, 224n82
Needham, Joseph, 107, 210n51
Newton, Alfred, 99, 145
New Year, Chinese, 29, 30, 129, 220n38
Notes and Queries on China & Japan (journal), 97, 101

Oiseaux de la Chine, Les (David), 105
Old Canton, times of, 15, 43, 104, 125, 126
Oldham, Richard, 126, 127, 128, 164
On the Origin of Species (Darwin), 71
Opium War, First (1839–1842), 4, 5, 14, 44; Britain's civilizing mission and, 23; British access to Chinese coastal ports and, 61; exploration and, 126; intensification of imperialism and, 64; Qing trade policy and, 6, 15; scientific explorations in the interior of China and, 125; Sino-Western contact following, 95–96
Opium War, Second (1856–1858), 61, 158, 188n42
Orientalism, 7, 64, 88, 115, 204n123; comparative philology and, 104; sinology as branch of, 92
Origin of Cultivated Plants (Candolle), 119
Ornithology, 74, 75, 111, 130, 145, 146
Orwell, George, 142
Otherness, discourse of, 147
Owen, Richard, 114

Paleontology, 118, 159
Parker, Edward H., 69, 148, 149, 164
Parks, John Damper, 24, 25, 26
Pearl River, 14, 16, 28
Peasants, 139, 149, 150
Performance, 137–138, 141–143
Petiver, James, 46
Pharmacology, 79, 80, 102
Philippines, 32
Philology, 100, 103, 112, 116
Pidgin, 16, 31, 35
Plantae Davidianae (Franchet), 105
Plants, 14, 34–35, 85, 125; cash crop plantations, 65; cataloging of, 57, 69; Chinese cultural symbolism and, 31; Chinese knowledge of, 83; in Chinese literature, 100, 105, 109; classification of, 111; commercial products from, 62; cultivated, 120; "discovery" of species, 89; drawings of, 42, 46–47, 49, 51–52; favorite Chinese species, 33; geographical distribution of, 102; introduced into China from abroad, 117; medicinal, 79–80; ornamental, 29, 62, 126, 146; sold in marketplace, 27; taxonomic nomenclature of, 71; transported overseas, 17–18, 19, 36–38. *See also* Botany
Playfair, G. M. H., 74, 164, 199n69
Plum tree, symbolism of, 54
Pompelly, Raphael, 126, 132
Porcelain workshops, 47–48
Portugal and the Portuguese, 44, 45
Potts, John, 25, 34, 36, 37
Power relations, 3, 4, 86, 138, 157
Pratt, Antwerp E., 131, 132, 133, 139, 140, 164
Protestant missionaries, 63, 73, 80–81, 97
Przhewalski, Nikolai, 123
Puankequa, 32
Puankequa II, 32, 33, 34, 35, 44

Index 237

Qing Empire, 6, 105; collapse of (1911), 6, 159; extent of authority of, 139–140; foreign colonial power and, 62–63; frontiers of, 151, 153; Hong merchants and, 32, 33; modernization campaign of, 158; policy on Western trade, 14–15; scholars and literati of, 50, 158; wars with Britain, 61; Western naturalists' travels under, 123. *See also* China

Raffles, Thomas Stamford, 57
Realism, 47, 51, 57
Reede tot Drakenstein, Hendrik Adriaan van, 12
Reeves, John, 21, 22, 23, 35; animal specimens obtained by, 29; biographical notes on, 164–165; Fa-tee nurseries and, 30; gardener-collectors and, 25; Hong merchants and, 34, 35; natural history drawings and, 41, 42–45, 49, 50, 51, 52; transporting of plants and, 36; visual authority and, 54–57
Reeves, John Russell, 22, 44, 165
Reptiles, 75, 130, 133
Ricci, Matteo, 94, 112
Richardson, John, 56
Richthofen, Ferdinand von, 82, 123, 126
Rickett, Charles Boughey, 82, 130, 131; biographical notes on, 165; collection of specimens and, 135, 136; folk knowledge and, 150
Riot, anti-Western (1883), 1, 169n1
Ross, John, 81, 165
Royal Asiatic Society, North China Branch (NCB), 82, 117, 119, 159; journal of, 113; Natural History Museum of Shanghai and, 115–116; orientalist scholarship and, 92; as Shanghai Literary and Scientific Society, 97
Royal Society, Fellows of the, 20, 21, 44
Rumphius, Georgius Everhardus, 12
Russia and the Russians, 14, 74, 123, 128, 192n84; imperial ambitions of, 153; rivalry with British Empire, 151

Sabine, Joseph, 55
Sailors, 17, 24, 27
Sampson, Theophilus, 1, 69, 79, 101, 104, 119; biographical notes on, 165; collection of specimens and, 134; plant names and, 117; on trustworthiness of Chinese texts, 115
Sancai tuhui (Chinese encyclopedia), 120
Sargent, Charles, 139

Science, 11, 35, 156; art and, 41–42; as gentlemen's hobby, 43; intellectual identity and, 99–100; "objective" knowledge and, 89–90; orientalist discourse and, 88; professionalization of, 91–92, 159, 206n1; scientific equipment, 142–143, 152; scientific networks, 75–80
Scientific imperialism, 4, 22, 62, 83–90; factual knowledge and, 83–89, 152; fieldwork and, 138. *See also* Imperialism
Sea captains: Biden, 36; Mayne, 19; M'Gilligan, 38; Wilson, 36, 37
Sedan chairs, naturalists carried in, 141–142, 225n107
Self-Strengthening movement, 158
Shanghai, 76, 77, 82, 129, 130, 202n105
Shanghai Literary and Scientific Society, 97
Shaw, Samuel, 48
Shen Fu, 29, 183n125
"Shooting an Elephant" (Orwell), 142
Shykinqua, 34
Siam, 151
Sichuan Province, 15, 62, 80, 125, 149; explorations in, 151; missionaries attacked in, 140
Sichuan tongzhi (Gazetteer of Sichuan), 110
Siebold, Philipp Franz von, 56, 105
Sinologists, 1, 44, 88; Chinese language and, 112, 113; Chinese texts and, 121; description of Chinese alligator and, 116, 118; knowledge translation and, 119; naturalists as, 97–99
Sinology, 6, 20, 92, 94–96, 100–104
Sino-Western relations, 2, 5
Slater, Gilbert, 19, 26, 46
Sloane, Hans, 46
Smith, Frederick Porter, 81, 102
Sowerby, Arthur de Carle, 159
Sowerby, James, 55
Specimens, 72, 74, 122; Chinese collectors of, 180n93; collection of, 12, 40, 79, 125, 131, 143; dried flowers, 46; folk knowledge and, 146; illustrations of, 41, 55; insects, 28; sabotage of, 138; scientific networks and, 21; scientific vocabulary and, 41
Spoilum (Chinese artist), 49
"Squire, the" (Hong merchant), 32, 44
Staunton, George Leonard, 20
Staunton, George Thomas, 20–21, 23, 43, 46, 96
Stevens, Thomas, 140
Strauss, David Friedrich, 72
Styan, Frederic William, 82, 130, 135, 136, 137, 165

Swinhoe, Robert, 69, 71, 74, 76, 114, 128; biographical notes on, 165; birds of Taiwan and, 146; on Chinese natural history illustration, 108; collection of specimens and, 135, 179n83; cultural encounter and, 141; fieldwork and, 144–145; folk knowledge and, 147, 148; journal articles by, 99; scientific networks of, 77, 100; sinology and, 98–99; tiger hunt and, 142; travels of, 123, 131; use of gazetteers, 110

Taintor, Edward, 79
Taiping Rebellion (1851–1864), 61, 129
Taiwan (Formosa), 75, 76, 99, 131, 155; aborigines of, 74; animals of, 78, 110; collection of specimens in, 136; naturalists' travels in, 125
Tang Wang-wang, 137
Tartary, 28
Taxidermy, 136, 137, 224n82
Taxonomy, 43, 71, 82, 89; in Chinese texts, 106, 118; fieldwork and, 144; folk knowledge and, 148; medicinal, 102
Tea culture, 126, 133–134, 143–144, 172n10
Technology, 153, 158
Teijismann, Johannes E., 79
Temple, William, 13
Textual practice, 111–115
Thiselton-Dyer, William, 70–71, 75, 77, 91, 109
Tibet, 28, 80, 123, 125, 132, 140
Tigers, 78, 142, 226n111
Tinqua, 48, 49
Tonqua, 49
Tourism, 129
Trader-naturalists, 17–23, 34
Translation: between knowledge traditions, 94, 120; strategies of, 112–113; and visual culture, 42; of Western science books, 158
Travel writing, 86, 123, 124, 151–152
Treaty ports, 76, 81, 138
Tropical medicine, 73

University of Aberdeen, 73
University of London, 73

Vacheli, George, 21, 22, 29, 45, 165
Variation of Animals and Plants under Domestication, The (Darwin), 120
Veitch, John, 126, 140, 156
Victorian culture, 74, 125, 128–129, 152
Visual authority, 54–57

Visual cultures, 42, 47–49
Vogt, Carl, 72

Wallace, Alfred Russel, 144, 147, 156
Wang Shu-hang, 137
Wang Tao, 94
Wardian case, 37
Watters, Thomas, 101
Webb, John, 94
Weber, Max, 31–32
Wenren hua (literati painting), 50
Whampoa (outer port of Canton), 16, 27, 68, 78
Wilford, Charles, 126, 127, 128
Wilkins, John, 95
Williams, S. Wells, 81, 88, 106
Williamson, Alexander, 81
Wilson, Ernest Henry, 62, 127, 131, 132, 159; biographical notes on, 165; relations with Chinese employees, 139
Wolley, John, 145
Women, 129, 197n50
Wu Qijun, 94, 109, 189n55, 212n68, 213n88

Xinjiang, 123, 153

Yangzhou huafang lu (Li Dou), 33
Yangzi River region, 61, 64, 76, 133, 140; birds in, 28; flowers from, 29–30; hunting expeditions in, 129–130; scientific exploration in, 126; trade routes and, 15
Yantai (Chefoo), 76, 129
Yichang, 76, 136, 140
Younghusband, Francis, 129, 142
Yuan Mei, 33
Yuan Ming Yuan, 13, 61
Yunnan Province, 62, 75, 125, 155; border with Burma, 139; border with Laos, 136; missionaries in, 80; plague in, 151; specimens collected in, 28, 131; trade routes through, 15

Zhang Dai, 29
Zhang Qian, 103, 212n64
Zhao Xuemin, 94
Zhiwu mingshi tukao (Wu Qijun), 94, 108–109, 189n55, 212n68
Zoological Society of London, 28, 44, 50, 83
Zoology, 76, 77, 81–82, 100, 130; hunting and, 125; scientific journals devoted to, 99